LOGIC FOR MATHEMATICIANS

LOGIC FOR MATHEMATICIANS

A. G. HAMILTON

Department of Computing Science, University of Stirling

Revised edition

CAMBRIDGE
UNIVERSITY PRESS

CAMBRIDGE
UNIVERSITY PRESS

Shaftesbury Road, Cambridge CB2 8EA, United Kingdom

One Liberty Plaza, 20th Floor, New York, NY 10006, USA

477 Williamstown Road, Port Melbourne, VIC 3207, Australia

314–321, 3rd Floor, Plot 3, Splendor Forum, Jasola District Centre, New Delhi – 110025, India

103 Penang Road, #05–06/07, Visioncrest Commercial, Singapore 238467

Cambridge University Press is part of Cambridge University Press & Assessment, a department of the University of Cambridge.

We share the University's mission to contribute to society through the pursuit of education, learning and research at the highest international levels of excellence.

www.cambridge.org
Information on this title: www.cambridge.org/9780521368650

First published 1978
Reprinted 1980, 1984
Revised edition 1988
Reprinted 1989, 1990, 1991, 1993, 1995, 1997, 2000

A catalogue record for this publication is available from the British Library

Library of Congress Cataloging-in-Publication data
Hamilton, Alan G., 1943–
Logic for mathematicians
Bibliography
Includes index
1. Logic, symbolic and mathematical. I. Title
QA.9.H298   511'.3   77–84802

ISBN   978-0-521-36865-0   Paperback

Contents

Preface

Every mathematician must know the conversation-stopping nature of the reply he gives to an inquiry by a non-mathematician about the nature of his business. For a logician in the company of other mathematicians to admit his calling is to invite similarly blank looks, admissions of ignorance, and a change in the topic of conversation. The rift between mathematicians and the public is a difficulty which will always exist (though no opportunity should be missed of narrowing it), but the rift between logicians and other mathematicians is, in my view, unnecessary. This book is an attempt to bridge the gap by providing an introduction to logic for mathematicians who do not necessarily aspire to becoming logicians.

Mathematical logic is now taught in many universities as part of an undergraduate course in mathematics or computing, and the subject is now coherent enough to have a standard body of fundamental material which must be included in any first course. This book is intended to be a textbook for such a course, but also to be something more – to be a book rather than merely a textbook. The material is deliberately presented in a direct manner, for its own sake, without particular bias towards any aspect, application or development of the subject. At the same time the attempt has been made to place the subject matter in the context of mathematics as a whole and to emphasise the relevance of logic to the mathematician.

The book is designed to be accessible to anyone with a mathematical background, from first year undergraduate to professional mathematician, who wishes, or is required, to find out something of what mathematical logic is. A certain familiarity with elementary algebra and number theory is assumed, and since ideas of countable and uncountable sets are fundamental, there is an appendix in which the necessary properties are described.

The material of the book is developed from that presented in two separate courses of sixteen lectures at the University of Stirling to students in their third and fourth years of undergraduate study. The first of these covered Chapters 1 to 4 with some of Chapter 5, and the second was a more advanced optional course covering the remainder. Chapter 6 is the most difficult in the book, but the significance of Gödel's Incompleteness Theorem is such that the ideas behind the proof ought to be brought out in a book of this kind. The detailed proofs may be omitted on first reading, as the material of Chapter 7 does not depend on them.

The scope of this book is more limited than that of other standard introductions to the subject. In particular the theory of models and the axiomatic theory of sets are barely touched on. The interested reader is therefore referred to the list at the end of the book of titles for further reading. Some of these are specifically referred to in the text (by the author's name), and overall they provide coverage of most areas of mathematical logic and treat in more depth the topics of this book.

There are exercises at the end of each section. Generally speaking, routine examples precede more taxing ones, but all the examples are intended as direct applications of the material in the corresponding section. Their purpose is to clarify and consolidate that material, not to extend it. Hints or solutions to many of the exercises are provided at the end of the book.

The symbols used in the book are, as far as is possible, standard (as is the terminology). There are some non-standard usages, however, which have been introduced in order to achieve clarity. These ought not to trouble the reader who is familiar with the material, and are intended to help the reader who is not. It is an unfortunate fact that different authors do use different notations and symbolism. For this reason, and for ease of reference, a glossary of symbols is included. Throughout the text the symbol $\triangleright$ is used to denote the resumption of the main exposition after it has been broken by a proposition, example, remark, corollary or definition.

Finally there are four debts which I wish to acknowledge. First, my debt to the book by Mendelson (*Introduction to Mathematical Logic*) will be apparent to all who are familiar with it. As a basic text for logicians it has had few rivals. Second, this book would not have been possible but for the time made available to me by the University of Stirling. Third, on a more personal level, I am most grateful to Francis Bell for his conscientious reading of a draft of the text and for his numerous valuable suggestions. And last, my sincere thanks go to Irene Wilson and May Abrahamson for all their patient labour in typing the manuscript.

1978   A. G. H.

1
Informal statement calculus

1.1 Statements and connectives

Logic, or at least logical mathematics, consists of deduction. We shall examine the rules of deduction making use of the precision which characterises a mathematical approach. In doing this, if we are to have any precision at all we must make our language unambiguous, and the standard mathematical way of doing that is to introduce a symbolic language, with the symbols having precisely stated meanings and uses. First of all we shall examine an aspect of everyday language, namely *connectives* (or conjunctions†, which is the more common grammatical term).

When we try to analyse a sentence in the English language, we can first note whether it is a simple sentence or a compound sentence. A simple sentence has a subject and a predicate (in the grammatical sense), for example:

> <u>Napoleon</u> is dead
> <u>John</u> owes James two pounds
> <u>All eggs which are not square</u> are round.

In each case the subject is underlined, and the remainder is the predi-cate. A compound sentence is made up from simple sentences by means of connectives, for example:

> Napoleon is dead and the world is rejoicing
> If all eggs are not square then all eggs are round
> If the barometer falls then either it will rain or it will snow.

We shall regard it as a basic assumption that all the simple sentences which we consider will be either true or false. It could certainly be argued that there are sentences which could not be regarded as either true or false, so we shall use a different term. We shall refer to simple and compound *statements*, and our assumption will be that all state-ments are either true or false.

---

† The word 'conjunction' has a more specific meaning for us. It is defined in Section 1.2.

Simple statements will be denoted by capital letters A, B, C, ... so in order to symbolise compound statements we have to introduce symbols for the connectives. The most common connectives, and the symbols which we shall use to denote them, are given in the table below.

not A	$\sim A$
A and B	$A \wedge B$
A or B	$A \vee B$
if A then B	$A \rightarrow B$
A if and only if B	$A \leftrightarrow B$

Of course, if the meaning of the symbols is to be defined precisely, we must be sure that we know precisely the meanings of the expressions in the left hand column. We shall return to this shortly.

The three compound statements above could be written in symbols (respectively) thus:

$$A \wedge B$$

$$C \rightarrow D$$

$$E \rightarrow (F \vee G)$$

where A stands for 'Napoleon is dead', B stands for 'the world is rejoicing', C stands for 'all eggs are not square', etc.

Notice that what remains when a compound statement is symbolised in this way is the bare logical bones, a mere 'statement form', which several different statements might have in common. It is precisely this which enables us to analyse deduction. For deduction has to do with the 'forms' of the statements in an argument rather than their meanings.

Example 1.1

> If Socrates is a man then Socrates is mortal
> Socrates is a man
> ∴ Socrates is mortal.

This is an argument which is regarded as logically satisfactory. But consider the argument:

> Socrates is a man
> ∴ Socrates is mortal.

The conclusion may be thought to follow from the premiss, but it does so because of the meanings of the words 'man' and 'mortal', not by a mere logical deduction. Let us put these arguments into symbols.

$$A \rightarrow B \qquad\qquad A$$
$$A \qquad\qquad\qquad \therefore\ B$$
$$\therefore\ B$$

It is the 'form' of the first which makes it valid. Any argument with the same form would also be valid. This is our logical intuition about if...then...statements. However, the second does not share this property. There are many arguments of this form which we would not regard intuitively as valid. For example:

The moon is yellow
∴   The moon is made of cheese.

We study, therefore, *statement forms* rather than particular statements. The letters $p, q, r, \ldots$ will be *statement variables* which stand for arbitrary and unspecified simple statements. Notice the distinction between the usages of the letters $p, q, r, \ldots$ and the letters $A, B, C, \ldots$ The former are variables for which particular simple statements may be substituted. The latter are merely 'labels' for particular simple statements. The variables enable us to describe in general the properties that statements and connectives have. Now each simple statement is either true or false, so a given statement variable can be thought of as taking one or other of the two *truth values*: T (true) or F (false). The way in which the truth or falsity of a compound statement or statement form depends on the truth or falsity of the simple statements or statement variables which constitute it is the subject of the next section.

Exercises

1   Translate into symbols the following compound statements.
  (a) If demand has remained constant and prices have been increased, then turnover must have decreased.
  (b) We shall win the election, provided that Jones is elected leader of the party.
  (c) If Jones is not elected leader of the party, then either Smith or Robinson will leave the cabinet, and we shall lose the election.
  (d) If x is a rational number and y is an integer, then z is not real.
  (e) Either the murderer has left the country or somebody is harbouring him.
  (f) If the murderer has not left the country, then somebody is harbouring him.
  (g) The sum of two numbers is even if and only if either both numbers are even or both numbers are odd.
  (h) If y is an integer then z is not real, provided that x is a rational number.
2   (a) Pick out any pairs of statements from the list in Exercise 1 which have the same form.
  (b) Pick out any pairs of statements from the list in Exercise 1 which have the same meaning.

1.2 Truth functions and truth tables

Let us consider the connectives in turn.

Negation

The negation of a statement A we write $\sim A$. Clearly if A is true then $\sim A$ is false, and if A is false then $\sim A$ is true. The meaning of A is irrelevant. We can describe the situation by a *truth table*:

p	$\sim p$
T	F
F	T

The table gives the truth value of $\sim p$, given the truth value of p. The connective $\sim$ gives rise to a *truth function*, $f^{\sim}$, in this case a function from the set $\{T, F\}$ to itself, given by the truth table, thus:

$$f^{\sim}(T) = F,$$
$$f^{\sim}(F) = T.$$

Conjunction

As above, it is easy to see that the truth value taken by the conjunction $A \wedge B$ of two statements A and B depends only on the truth value taken by A and the truth value taken by B. We have the table:

p	q	$p \wedge q$
T	T	T
T	F	F
F	T	F
F	F	F

We have in the table one row for each of the possible combinations of truth values for p and q. The last column gives the corresponding truth values for $p \wedge q$. The connective $\wedge$ thus defines a truth function $f^{\wedge}$ of two places.

$$f^{\wedge}(T, T) = T,$$
$$f^{\wedge}(T, F) = F,$$
$$f^{\wedge}(F, T) = F,$$
$$f^{\wedge}(F, F) = F.$$

Disjunction

We have used $A \vee B$ to denote 'A or B,' but there are two distinct standard usages of the word 'or' in English. 'A or B' may mean 'A or B or both', or it may mean 'A or B but not both'. In order to keep our symbolic language precise we must choose only one of these to give the meaning of our symbol $\vee$. We choose the former. There is no particular reason for this; we could just as well have chosen the latter. The truth table is as follows:

p	q	$p \vee q$
T	T	T
T	F	T
F	T	T
F	F	F

The connective $\vee$ defines a truth function of two places just as $\wedge$ did.

Remark. If A and B are simple statements, we can symbolise 'A or B but not both' as

$$(A \vee B) \wedge \sim (A \wedge B).$$

Correspondingly, if we had used 'A or B but not both' to define our disjunction symbol, we could have expressed 'A or B or both' using that disjunction along with $\wedge$ and $\sim$.

Conditional

$A \to B$ is to represent the statement 'A implies B' or 'if A then B'. Now in this case normal English usage is not as helpful in constructing a truth table, and the table that we use is a common source of intuitive difficulty. It is:

p	q	$p \to q$
T	T	T
T	F	F
F	T	T
F	F	T

The difficulty arises with the truth value T assigned to $A \to B$ in the cases where A is false. Consideration of examples of conditional statements in which the antecedent is false might perhaps lead one to the conclusion that such statements do not have a truth value at all. One might also gain the impression that such statements are not useful or meaningful. For example, the statement:

> If grass is red then the moon is made of green cheese

could fairly be said to be meaningless.

However, we shall be interested in deduction and methods of proof, principally in mathematics. In this context the significance of a conditional statement $A \rightarrow B$ is that its truth enables the truth of B to be inferred from the truth of A, and nothing in particular to be inferred from the falsity of A. A very common sort of mathematical statement can serve to illustrate this, namely a *universal* statement, for example:

> For every integer n, if $n > 2$ then $n^2 > 4$.

This is regarded as a true statement about integers. We would expect, therefore, to regard the statement

> If $n > 2$ then $n^2 > 4$

as true, irrespective of the value taken by n. Different values of n give rise to all possible combinations of truth values for '$n > 2$' and '$n^2 > 4$' except the combination *TF*. Taking n to be 3, -3, 1 respectively yields the combinations *TT*, *FT*, *FF*, and these are the combinations which, according to our truth table, give the implication the truth value *T*. The intuitive truth of this implication is therefore some justification for the truth table. The point to remember is that the only circumstance in which the statement $A \rightarrow B$ is regarded as false is when A is true and B is false.

Biconditional

We denote 'A if and only if B' by $A \leftrightarrow B$. The situation here is clear. We should have $A \leftrightarrow B$ true when and only when A and B have the same truth value (both true or both false). The truth table is then as shown.

p	q	$p \leftrightarrow q$
T	T	T
T	F	F
F	T	F
F	F	T

This completes our list of connectives. Obviously, compound statements of any length can be built up from simple statements using these connectives. Using statement variables we can build up statement *forms* of any length.

Definition 1.2

A *statement form* is an expression involving statement variables and connectives, which can be formed using the rules:
  (i) Any statement variable is a statement form.
  (ii) If $\mathscr{A}$ and $\mathscr{B}$ are statement forms, then $(\sim\mathscr{A})$, $(\mathscr{A} \wedge \mathscr{B})$, $(\mathscr{A} \vee \mathscr{B})$, $(\mathscr{A} \to \mathscr{B})$, and $(\mathscr{A} \leftrightarrow \mathscr{B})$ are statement forms.

Example 1.3

$((p \wedge q) \to (\sim(q \vee r)))$ is a statement form. By (i), p, q, r are statement forms. By (ii), $(p \wedge q)$ and $(q \vee r)$ are statement forms. By (ii), $(\sim(q \vee r))$ is a statement form. By (ii), $((p \wedge q) \to (\sim(q \vee r)))$ is a statement form.

▷ This definition is an example of an inductive definition. It sets a pattern which will occur again when we describe formal systems in detail.

The connectives determine simple truth functions. Using the truth tables for the connectives, we can construct a truth table for any given statement form. By this is meant a table which will indicate, for any given assignment of truth values to the statement variables appearing in the statement form, the truth value which it takes. This truth table is a graphical representation of a truth function. Thus each statement form gives rise to a truth function, the number of arguments of the function being the number of different statement variables appearing in the statement form. Let us illustrate this by means of some examples.

Example 1.4

  (a) $((\sim p) \vee q)$.

First construct the truth table:

p	q	$(\sim p)$	$((\sim p) \vee q)$
T	T	F	T
T	F	F	F
F	T	T	T
F	F	T	T

Observe that the truth function corresponding to this statement form is the same as the truth function determined by $(p \to q)$.

  (b) $(p \to (q \vee r))$.

Truth table:

p	q	r	$(q \vee r)$	$(p \to (q \vee r))$
T	T	T	T	T
T	T	F	T	T
T	F	T	T	T
T	F	F	F	F
F	T	T	T	T
F	T	F	T	T
F	F	T	T	T
F	F	F	F	T

The truth function here is a three place function, since there are three statement variables. Each row of the table gives the value of the truth function for a different combination of truth values for the letters. Notice that there will be eight rows in the truth table of any statement form involving three statement variables, and notice the pattern in which the first three columns of the table above are written out. This way of grouping the Ts and Fs under the p, q, r ensures that each possible combination appears once and only once.

▷ In the general case, for a statement form involving n different statement variables (n any natural number), the truth function will be a function of n places, and the truth table will have 2^n rows, one for each of the possible combinations of truth values for the statement variables. Further, notice that there are 2^{2^n} distinct truth functions with n places, corresponding to the 2^{2^n} possible ways of arranging the Ts and Fs in the last column of a truth table with 2^n rows. The number of statement forms which can be constructed using n statement variables is clearly infinite, so it follows that different statement forms may correspond to the same truth function.

To investigate this further we need some definitions.

Definition 1.5

(a) A statement form is a *tautology* if it takes truth value T under each possible assignment of truth values to the statement variables which occur in it.

(b) A statement form is a *contradiction* if it takes truth value F under each possible assignment of truth values to the statement variables which occur in it.

▷ Not every statement form falls into one or other of these categories. In fact none of those which we have considered so far does.

Example 1.6

(a) $(p \lor (\sim p))$ is a tautology.
(b) $(p \land (\sim p))$ is a contradiction.
(c) $(p \leftrightarrow (\land (\sim p)))$ is a tautology.
(d) $(((\sim p) \to q) \to (((\sim p) \to (\sim q)) \to p))$ is a tautology.

The method used to verify that a given statement form is either a tautology or a contradiction is just construction of the truth table.

▷ It should be clear from the definition that all tautologies containing n statement variables give rise to the same truth function of n places, namely that which takes value T always. We can make a similar observation about contradictions.

Definition 1.7

If $\mathcal{A}$ and $\mathcal{B}$ are statement forms, $\mathcal{A}$ *logically implies* $\mathcal{B}$ if $(\mathcal{A} \to \mathcal{B})$ is a tautology, and $\mathcal{A}$ is *logically equivalent* to $\mathcal{B}$ if $(\mathcal{A} \leftrightarrow \mathcal{B})$ is a tautology.

Example 1.8

(a) $(p \land q)$ logically implies p.
(b) $(\sim(p \land q))$ is logically equivalent to $((\sim p) \lor (\sim q))$.
(c) $(\sim(p \lor q))$ is logically equivalent to $((\sim p) \land (\sim q))$.

For (a): truth table of $((p \land q) \to p)$:

((p	$\land$	q)	$\to$	p)
T	T	T	T	T
T	F	F	T	T
F	F	T	T	F
F	F	F	T	F

For (b):

($\sim$	(p	$\land$	q)	$\leftrightarrow$	(($\sim$	p)	$\lor$	($\sim$	q)))
F	T	T	T	T	F	T	F	F	T
T	T	F	F	T	F	T	T	T	F
T	F	F	T	T	T	F	T	F	T
T	F	F	F	T	T	F	T	T	F

Here we have introduced a different way of writing the truth tables. For complicated statement forms it is easier to construct the table this way. Start by writing columns of Ts and Fs under the statement variables in the same order as we have in previous tables, to ensure that each combination appears once only. This must be done consistently throughout, of course. Next, under the connectives successively insert the truth values of the parts, until the column giving the truth values of the whole statement form is filled up. This column is enclosed by vertical lines in the examples above.

Remark. Let $\mathscr{A}$ and $\mathscr{B}$ be statement forms containing the same statement variables. If $\mathscr{A}$ and $\mathscr{B}$ are logically equivalent then they represent the same truth function. For if $(\mathscr{A} \leftrightarrow \mathscr{B})$ is a tautology it never takes value F, and so $\mathscr{A}$ and $\mathscr{B}$ must always take the same truth value. The truth functions corresponding to $\mathscr{A}$ and $\mathscr{B}$ must therefore be the same.

Exercises

3   Write out the truth tables of the following statement forms:

(a) $((\sim p) \wedge (\sim q))$;
(b) $\sim((p \rightarrow q) \rightarrow (\sim(q \rightarrow p)))$;
(c) $(p \rightarrow (q \rightarrow r))$;
(d) $((p \wedge q) \rightarrow r)$;
(e) $((p \leftrightarrow (\sim q)) \vee q)$;
(f) $((p \wedge q) \vee (r \wedge s))$;
(g) $(((\sim p) \wedge q) \rightarrow ((\sim q) \wedge r))$;
(h) $((p \rightarrow (q \rightarrow r)) \rightarrow ((p \rightarrow q) \rightarrow (p \rightarrow r)))$.

4   Show that the statement form $((\sim p) \vee q)$ gives rise to the same truth function as $(p \rightarrow q)$, and that $((\sim p) \rightarrow (q \vee r))$ gives rise to the same truth function as $((\sim q) \rightarrow ((\sim r) \rightarrow p))$.

5   Which of the following statement forms are tautologies?

(a) $(p \rightarrow (q \rightarrow p))$;
(b) $((q \vee r) \rightarrow ((\sim r) \rightarrow q))$;
(c) $((p \wedge (\sim q)) \vee ((q \wedge (\sim r)) \vee (r \wedge (\sim p))))$;
(d) $((p \rightarrow (q \rightarrow r)) \rightarrow ((p \wedge (\sim q)) \vee r))$.

6   Show that the following pairs of statement forms are logically equivalent.

(a) $(p \rightarrow q), ((\sim q) \rightarrow (\sim p))$;
(b) $((p \vee q) \wedge r), ((p \wedge r) \vee (q \wedge r))$;
(c) $(((\sim p) \wedge (\sim q)) \rightarrow (\sim r)), (r \rightarrow (q \vee p))$;
(d) $(((\sim p) \vee q) \rightarrow r), ((p \wedge (\sim q)) \vee r)$.

7   Show that the statement form $(((\sim p) \rightarrow q) \rightarrow (p \rightarrow (\sim q)))$ is not a tautology. Find statement forms $\mathscr{A}$ and $\mathscr{B}$ such that $(((\sim \mathscr{A}) \rightarrow \mathscr{B}) \rightarrow (\mathscr{A} \rightarrow (\sim \mathscr{B})))$ is a contradiction.

1.3 Rules for manipulation and substitution

Proposition 1.9

If $\mathscr{A}$ and $(\mathscr{A} \rightarrow \mathscr{B})$ are tautologies, then $\mathscr{B}$ is a tautology.

Proof. Suppose that $\mathcal{A}$ and $(\mathcal{A} \to \mathcal{B})$ are tautologies, and that $\mathcal{B}$ is not. Then there is an assignment of truth values to the statement variables appearing in $\mathcal{A}$ or in $\mathcal{B}$ which gives $\mathcal{B}$ the value F. But it must give $\mathcal{A}$ the value T since $\mathcal{A}$ is a tautology, and so it gives $(\mathcal{A} \to \mathcal{B})$ the value F. This contradicts the assumption that $(\mathcal{A} \to \mathcal{B})$ is a tautology. Hence $\mathcal{B}$ must be a tautology.

▷ Consider the statement form $(p \to p)$. It is easy to show that this is a tautology. Now if we substitute the statement form $((r \wedge s) \to t)$ for p in both places, we obtain

$$(((r \wedge s) \to t) \to ((r \wedge s) \to t)),$$

which is again a tautology. It is intuitively clear that this would happen whatever statement form we substituted for p, as long as we substituted it for *every* occurrence of p. (The statement form $(p \to ((r \wedge s) \to t))$ is obviously not a tautology.) We collect this idea into a proposition.

Proposition 1.10

Let $\mathcal{A}$ be a statement form in which the statement variables $p_1, p_2, \ldots, p_n$ appear, and let $\mathcal{A}_1, \mathcal{A}_2, \ldots, \mathcal{A}_n$ be any statement forms. If $\mathcal{A}$ is a tautology then the statement form $\mathcal{B}$, obtained from $\mathcal{A}$ by replacing each occurrence of p_i by $\mathcal{A}_i$ $(1 \le i \le n)$ throughout, is a tautology also.

Proof. Let $\mathcal{A}$ be a tautology and let $p_1, p_2, \ldots, p_n$ be the statement variables appearing in $\mathcal{A}$. Let $\mathcal{A}_1, \mathcal{A}_2, \ldots, \mathcal{A}_n$ be any statement forms. Assign any truth values to the statement variables which appear in $\mathcal{A}_1, \mathcal{A}_2, \ldots, \mathcal{A}_n$. The truth value that $\mathcal{B}$ now takes is the same as that which $\mathcal{A}$ would have taken if the values which $\mathcal{A}_1, \mathcal{A}_2, \ldots, \mathcal{A}_n$ take had been assigned to $p_1, p_2, \ldots, p_n$ respectively, namely T. Hence $\mathcal{B}$ takes value T under any assignment of truth values, i.e. $\mathcal{B}$ is a tautology.

▷ Proposition 1.10 is one of several we shall come across whose applications are widespread and very often unconscious. As an example, the next proposition is a useful result.

Proposition 1.11

For any statement forms $\mathcal{A}$ and $\mathcal{B}$, $(\sim(\mathcal{A} \wedge \mathcal{B}))$ is logically equivalent to $((\sim\mathcal{A}) \vee (\sim\mathcal{B}))$, and $(\sim(\mathcal{A} \vee \mathcal{B}))$ is logically equivalent to $((\sim\mathcal{A}) \wedge (\sim\mathcal{B}))$.

Proof. We saw previously that

$$(\sim(p \wedge q) \leftrightarrow ((\sim p) \vee (\sim q)))$$

is a tautology. It is a consequence of Proposition 1.10 that for any statement forms $\mathcal{A}$ and $\mathcal{B}$,

$$((\sim(\mathcal{A} \wedge \mathcal{B})) \leftrightarrow ((\sim\mathcal{A}) \vee (\sim\mathcal{B})))$$

is also a tautology. Hence $(\sim(\mathcal{A} \wedge \mathcal{B}))$ is logically equivalent to $((\sim\mathcal{A}) \vee (\sim\mathcal{B}))$. The other part is proved similarly.

Example 1.12

For any statement forms $\mathcal{A}$, $\mathcal{B}$, $\mathcal{C}$, $(\mathcal{A} \wedge (\mathcal{B} \wedge \mathcal{C}))$ is logically equivalent to $((\mathcal{A} \wedge \mathcal{B}) \wedge \mathcal{C})$. (Because of this we customarily omit the inner parentheses, and write $(\mathcal{A} \wedge \mathcal{B} \wedge \mathcal{C})$.)

Consider the statement form

$$(((p_1 \wedge (p_2 \wedge p_3)) \leftrightarrow ((p_1 \wedge p_2) \wedge p_3))).$$

By constructing a truth table in the usual way it can be shown that this is a tautology. Now apply Proposition 1.10, substituting $\mathcal{A}$, $\mathcal{B}$, $\mathcal{C}$ for p_1, p_2, p_3 respectively, to obtain the desired result.

Example 1.13

For any statement forms $\mathcal{A}$, $\mathcal{B}$, $\mathcal{C}$, the following pairs of statement forms are logically equivalent.

(a) $(\mathcal{A} \vee (\mathcal{B} \vee \mathcal{C}))$ and $((\mathcal{A} \vee \mathcal{B}) \vee \mathcal{C})$.
(b) $(\mathcal{A} \wedge \mathcal{B})$ and $(\mathcal{B} \wedge \mathcal{A})$.
(c) $(\mathcal{A} \vee \mathcal{B})$ and $(\mathcal{B} \vee \mathcal{A})$.

The verifications of these follow the same pattern as the worked example above.

▷ Now consider the statement form $((p \wedge p) \rightarrow q)$. $(p \wedge p)$, which appears in this form, is logically equivalent to p (since $((p \wedge p) \leftrightarrow p)$ is a tautology). If we replace $(p \wedge p)$ by p, we get $(p \rightarrow q)$. Now $(p \rightarrow q)$ is logically equivalent to $((p \wedge p) \rightarrow q)$ (check by truth table). Again this is an instance of a general proposition about substitution.

Proposition 1.14

If $\mathcal{B}_1$ is a statement form arising from the statement form $\mathcal{A}_1$ by substituting the statement form $\mathcal{B}$ for one or more occurrences of the statement form $\mathcal{A}$ in $\mathcal{A}_1$, and if $\mathcal{B}$ is logically equivalent to $\mathcal{A}$, then $\mathcal{B}_1$ is logically equivalent to $\mathcal{A}_1$.

Proof. Suppose that $\mathcal{B}$ is logically equivalent to $\mathcal{A}$ and that $\mathcal{B}_1$ and $\mathcal{A}_1$ are as described. We wish to show that $(\mathcal{A}_1 \leftrightarrow \mathcal{B}_1)$ is a tautology. Assign truth values to all the statement variables involved. $\mathcal{B}_1$ differs from $\mathcal{A}_1$

only in having $\mathscr{B}$ in some places where $\mathscr{A}$ was. The truth value that $\mathscr{B}_1$ takes must be the same as that taken by $\mathscr{A}_1$ since $\mathscr{A}$ and $\mathscr{B}$ have the same truth value. Therefore $(\mathscr{A}_1 \leftrightarrow \mathscr{B}_1)$ takes value T. It follows that $(\mathscr{A}_1 \leftrightarrow \mathscr{B}_1)$ must always take value T, since the truth values originally assigned to the variables were arbitrary. Thus $(\mathscr{A}_1 \leftrightarrow \mathscr{B}_1)$ is a tautology and $\mathscr{A}_1$ is logically equivalent to $\mathscr{B}_1$.

▷ The next proposition is not an integral part of the development, but we include it here since it gives an impression of the methods which we shall be using later on. For this reason we shall go into the proof in some detail.

Proposition 1.15

Let us refer to a statement form involving only the connectives $\sim$, $\wedge$ and $\vee$ as a restricted statement form. Let $\mathscr{A}$ be a restricted statement form and suppose that $\mathscr{A}^*$ results from $\mathscr{A}$ by interchanging $\wedge$ and $\vee$ and replacing every statement variable by its negation, throughout $\mathscr{A}$. Then $\mathscr{A}^*$ is logically equivalent to $(\sim \mathscr{A})$.

Proof. The proof is by induction on the number n of connectives which appear in $\mathscr{A}$. If we can prove that, for each natural number n, every restricted statement form $\mathscr{A}$ with exactly n connectives satisfies the proposition, then clearly that will be sufficient.

Base Step: $n = 0$ ($\mathscr{A}$ contains no connectives). In this case $\mathscr{A}$ consists merely of a statement variable, p, say. So here $\mathscr{A}^*$ is $(\sim p)$, so, trivially, $\mathscr{A}^*$ is logically equivalent to $(\sim \mathscr{A})$.

Induction Step: Suppose that $n > 0$, that $\mathscr{A}$ has n connectives, and that every restricted statement form with fewer than n connectives has the required property. Because of the ways that statement forms can be built up, we have to consider three cases:

Case 1: 　$\mathscr{A}$ has the form $(\sim \mathscr{B})$.
Case 2: 　$\mathscr{A}$ has the form $(\mathscr{B} \vee \mathscr{C})$.
Case 3: 　$\mathscr{A}$ has the form $(\mathscr{B} \wedge \mathscr{C})$.

For Case 1: $\mathscr{B}$ has $n - 1$ connectives, so by the hypothesis of induction $\mathscr{B}^*$ is logically equivalent to $(\sim \mathscr{B})$. But $\mathscr{A}^*$ is $(\sim \mathscr{B}^*)$, so $\mathscr{A}^*$ is logically equivalent to $(\sim(\sim \mathscr{B}))$, i.e. to $(\sim \mathscr{A})$. Notice that, almost unawares, we have used the result of Proposition 1.14 here.

For Case 2: $\mathscr{B}$ and $\mathscr{C}$ each contain fewer than n connectives, so $\mathscr{B}^*$ and $\mathscr{C}^*$ are logically equivalent to $(\sim \mathscr{B})$ and $(\sim \mathscr{C})$ respectively. Now $\mathscr{A}^*$ is $(\mathscr{B}^* \wedge \mathscr{C}^*)$. By Proposition 1.14 this is logically equivalent to $((\sim \mathscr{B}) \wedge \mathscr{C}^*)$ and by the same proposition again this is logically equivalent to

$((\sim\mathscr{B}) \wedge (\sim\mathscr{C}))$. Now we have seen (Proposition 1.11) that this is logically equivalent to $(\sim(\mathscr{B} \vee \mathscr{C}))$, i.e. $(\sim\mathscr{A})$. Thus $\mathscr{A}^*$ is logically equivalent to $(\sim\mathscr{A})$.

For Case 3: As in Case 2, $\mathscr{B}^*$ and $\mathscr{C}^*$ are logically equivalent to $(\sim\mathscr{B})$ and $(\sim\mathscr{C})$ respectively. $\mathscr{A}^*$ is $(\mathscr{B}^* \vee \mathscr{C}^*)$, which is logically equivalent to $((\sim\mathscr{B}) \vee \mathscr{C}^*)$, and so to $((\sim\mathscr{B}) \vee (\sim\mathscr{C}))$, and so to $(\sim(\mathscr{B} \wedge \mathscr{C}))$, i.e. to $(\sim\mathscr{A})$. Thus $\mathscr{A}^*$ is logically equivalent to $(\sim\mathscr{A})$.

Under the assumption that every restricted statement form with fewer than n connectives has the required property, we have shown that every restricted statement form with n connectives has the required property. By the principle of mathematical induction then, every restricted statement form has the required property.

Corollary 1.16

If $p_1, p_2, \ldots, p_n$ are statement variables, then

$$((\sim p_1) \vee (\sim p_2) \vee \ldots \vee (\sim p_n))$$

is logically equivalent to

$$(\sim(p_1 \wedge p_2 \wedge \ldots \wedge p_n)).$$

Proof. This is a special case of Proposition 1.15, in which $\mathscr{A}$ is the statement form $(p_1 \wedge p_2 \wedge \ldots \wedge p_n)$.

▷ Introducing a new notation, for the sake of brevity, we can write this result as

$$\left(\bigvee_{i=1}^{n} (\sim p_i)\right) \text{ is logically equivalent to } \left(\sim\left(\bigwedge_{i=1}^{n} p_i\right)\right).$$

We can use the proposition also to prove the 'dual' of this, namely

$$((\sim p_1) \wedge (\sim p_2) \wedge \ldots \wedge (\sim p_n))$$

is logically equivalent to

$$\sim(p_1 \vee p_2 \vee \ldots \vee p_n),$$

i.e. $\left(\bigwedge_{i=1}^{n} (\sim p_i)\right)$ is logically equivalent to $\left(\sim\left(\bigvee_{i=1}^{n} p_i\right)\right).$

Proposition 1.17 (De Morgan's Laws)

Let $\mathscr{A}_1, \mathscr{A}_2, \ldots, \mathscr{A}_n$ be any statement forms. Then:

   (i) $(\bigvee_{i=1}^{n} (\sim\mathscr{A}_i))$ is logically equivalent to $(\sim(\bigwedge_{i=1}^{n} \mathscr{A}_i))$.
   (ii) $(\bigwedge_{i=1}^{n} (\sim\mathscr{A}_i))$ is logically equivalent to $(\sim(\bigvee_{i=1}^{n} \mathscr{A}_i))$.

Proof. Using the above corollaries and Proposition 1.10.

Exercises

8 Show that, for any statement forms $\mathcal{A}$, $\mathcal{B}$, $\mathcal{C}$, the following pairs of statement forms are logically equivalent.

   (a) $(\mathcal{A} \vee (\mathcal{B} \vee \mathcal{C}))$ and $((\mathcal{A} \vee \mathcal{B}) \vee \mathcal{C})$;
   (b) $(\mathcal{A} \wedge \mathcal{B})$ and $(\mathcal{B} \wedge \mathcal{A})$;
   (c) $(\mathcal{A} \vee \mathcal{B})$ and $(\mathcal{B} \vee \mathcal{A})$;
   (d) $\mathcal{A}$ and $(\sim(\sim\mathcal{A}))$.

9 Show that, for any statement forms $\mathcal{A}$ and $\mathcal{B}$, the following statement forms
   are tautologies.

   (a) $((\mathcal{A} \wedge \mathcal{B}) \to \mathcal{A})$;
   (b) $((\mathcal{A} \wedge \mathcal{B}) \to \mathcal{B})$.

10 Prove, using Proposition 1.14, that $((\sim((\sim p) \vee q)) \vee r)$ is logically equivalent
    to $((p \to q) \to r)$.

11 Show, using Propositions 1.14 and 1.17, that the statement form $((\sim(p \vee
    (\sim q))) \to (q \to r))$ is logically equivalent to each of the following.

    (a) $((\sim(q \to p)) \to ((\sim q) \vee r))$;
    (b) $(((\sim p) \wedge q) \to (\sim(q \wedge (\sim r))))$;
    (c) $((\sim((\sim q) \vee r)) \to (q \to p))$;
    (d) $(q \to (p \vee r))$.

1.4 Normal forms

Let us consider further what we have called restricted statement forms.
We have observed previously that from each statement form we can
construct a truth table. Now we prove a converse to that.

Proposition 1.18

Every truth function is the truth function determined by a statement
form in which the only connectives occurring are from amongst $\sim$, $\wedge$
and $\vee$ (i.e. a restricted statement form).

Proof. Suppose that the given function is a function of n places. We
shall construct a statement form $\mathcal{A}$ from the variables $p_1, p_2, \ldots, p_n$.
Notice first that if the truth function takes value F for each combination
of truth values, then it corresponds to any contradiction, and so the
statement form

$$((p_1 \wedge (\sim p_1)) \wedge p_2 \wedge p_3 \wedge \ldots p_n)$$

will do.

Now suppose that the truth function has value T at least once. Our
method depends on constructing, for each of the 2^n combinations of
truth values of the statement variables, a statement form which is true
for that combination but false for all other combinations. For example,
if $n = 3$, the statement form $(p_1 \wedge (\sim p_2) \wedge (\sim p_3))$ is true only for the

combination *TFF* of truth values of p_1, p_2, p_3 respectively, and $((\sim p_1) \wedge (\sim p_2) \wedge p_3)$ is true only for the combination *FFT*. These special statement forms are called *basic conjunctions*. Given an assignment of truth values to $p_1, p_2, \ldots, p_n$, we put p_i in the conjunction if p_i is assigned a *T*, and we put $(\sim p_i)$ in the conjunction if p_i is assigned an *F*, for $1 \le i \le n$. Then for the given assignment of truth values, each conjunct has value *T*, and so the whole conjunction has value *T*. Correspondingly, for any other assignment of truth values, at least one of the conjuncts will have value *F*, so that the whole conjunction will have value *F*.

Now for the proof of our proposition, consider all the combinations of *n* truth values for which our truth function yields value *T*. Let $\mathscr{A}$ be the disjunction of all the basic conjunctions obtained by taking these combinations as truth values of $p_1, p_2, \ldots, p_n$. This $\mathscr{A}$ is the required statement form. To see this assign truth values to $p_1, p_2, \ldots, p_n$. If our truth function applied to this combination yields a *T*, then the corresponding basic conjunction is included in $\mathscr{A}$ and has value *T* for this assignment of truth values, so $\mathscr{A}$ has value *T* also. If our truth function applied to this combination yields an *F*, then the corresponding basic conjunction is not included in $\mathscr{A}$, and each of the other basic conjunctions which are included in $\mathscr{A}$ takes value *F* for this assignment of truth values, so $\mathscr{A}$ has value *F* also. Thus for any assignment of truth values, the truth value of $\mathscr{A}$ is as given by the truth function.

▷ This proof is best understood by thinking of it in relation to a concrete example.

Example 1.19

Let us specify a truth function by means of a table (this is a three place function).

T	*T*	*T*	*T*
T	*T*	*F*	*T*
T	*F*	*T*	*F*
T	*F*	*F*	*F*
F	*T*	*T*	*F*
F	*T*	*F*	*F*
F	*F*	*T*	*F*
F	*F*	*F*	*T*

The combinations of truth values for which the function yields value *T*

are *TTT*, *TTF*, and *FFF*. The basic conjunctions corresponding to these
are:

$$(p_1 \wedge p_2 \wedge p_3),$$

$$(p_1 \wedge p_2 \wedge (\sim p_3)),$$

$$((\sim p_1) \wedge (\sim p_2) \wedge (\sim p_3)).$$

The statement form $\mathscr{A}$ constructed in the proof is

$$(p_1 \wedge p_2 \wedge p_3) \vee (p_1 \wedge p_2 \wedge (\sim p_3)) \vee ((\sim p_1) \wedge (\sim p_2) \wedge (\sim p_3)).$$

This statement form corresponds to the given truth function, and the
table given is the truth table of this statement form.

Corollary 1.20

Every statement form which is not a contradiction is logically equivalent
to a restricted statement form of the form $(\bigvee_{i=1}^{m}(\bigwedge_{j=1}^{n} Q_{ij}))$, where each
Q_{ij} is either a statement variable or the negation of a statement variable.
This form is called a *disjunctive normal form*.

Proof. Two statement forms are logically equivalent if and only if they
correspond to the same truth function. Given a statement form $\mathscr{A}$,
obtain its truth table and the truth function it defines. Then apply the
method of the proof of Proposition 1.18 to obtain a statement form in
the desired form corresponding to this truth function.

Corollary 1.21

Every statement form which is not a tautology is logically equivalent to a
restricted statement form of the form $(\bigwedge_{i=1}^{m} \bigvee_{j=1}^{n} Q_{ij}))$, where each Q_{ij} is
either a statement variable or the negation of a statement variable. This
form is called *conjunctive normal form*.

Proof. Let $\mathscr{A}$ be a statement form which is not a tautology. Then $(\sim\mathscr{A})$
is not a contradiction, and is logically equivalent to a statement form
$(\bigvee_{i=1}^{m} \bigwedge_{j=1}^{n} Q_{ij}))$ in disjunctive normal form. $\mathscr{A}$ is therefore logically
equivalent to $\sim(\bigvee_{i=1}^{m} (\bigwedge_{j=1}^{n} Q_{ij}))$, and, using De Morgan's laws, this is
logically equivalent to $(\bigwedge_{i=1}^{m} (\bigvee_{j=1}^{n} (\sim Q_{ij})))$. The result now follows
if we replace all expressions of the form $(\sim(\sim q))$ in this statement
form by q.

Example 1.22

Find a conjunctive normal form which is logically equivalent to

$$(((\sim p_1) \vee p_2) \rightarrow p_3).$$

First find the truth table of the negation of this.

~	(((~	p₁)	∨	p₂)	→	p₃)
F	F	T	T	T	T	T
T	F	T	T	T	F	F
F	F	T	F	F	T	T
F	F	T	F	F	T	F
F	T	F	T	T	T	T
T	T	F	T	T	F	F
F	T	F	T	F	T	T
T	T	F	T	F	F	F

The combinations which give value T are TTF, FTF, and FFF. So a disjunctive normal form which is logically equivalent is

$$((p_1 \wedge p_2 \wedge (\sim p_3)) \vee ((\sim p_1) \wedge p_2 \wedge (\sim p_3)) \vee$$
$$((\sim p_1) \wedge (\sim p_2) \wedge (\sim p_3))).$$

Hence the given statement form is logically equivalent to the negation of this, which by De Morgan's laws is logically equivalent to

$$(\sim (p_1 \wedge p_2 \wedge (\sim p_3)) \wedge \sim ((\sim p_1) \wedge p_2 \wedge (\sim p_3)) \wedge$$
$$\sim ((\sim p_1) \wedge (\sim p_2) \wedge (\sim p_3))),$$

and to

$$(((\sim p_1) \vee (\sim p_2) \vee (\sim (\sim p_3))) \wedge ((\sim (\sim p_1) \vee (\sim p_2) \vee (\sim (\sim p_3))) \wedge$$
$$((\sim (\sim p_1) \vee (\sim (\sim p_2)) \vee (\sim (\sim p_3))),$$

and to

$$((\sim p_1) \vee (\sim p_2) \vee p_3) \wedge (p_1 \vee (\sim p_2) \vee p_3) \wedge (p_1 \vee p_2 \vee p_3),$$

which is in conjunctive normal form.

Exercises

12 Find statement forms in disjunctive normal form which are logically equivalent to the following:

(a) $(p \leftrightarrow q)$;

(b) $(p \rightarrow ((\sim q) \vee r))$;

(c) $((p \wedge q) \vee ((\sim q) \leftrightarrow r))$;

(d) $\sim ((p \rightarrow (\sim q)) \rightarrow r)$;

(e) $(((p \rightarrow q) \rightarrow r) \rightarrow s)$.

13 Find statement forms in conjunctive normal form which are logically equivalent to the following:

(a) $(((\sim p) \vee q) \rightarrow r)$;
(b) $(p \leftrightarrow q)$;
(c) $(p \wedge q \wedge r) \vee ((\sim p) \wedge (\sim q) \wedge r)$;
(d) $(((p \rightarrow q) \rightarrow r) \rightarrow s)$.

1.5 Adequate sets of connectives

Definition 1.23

An *adequate* set of connectives is a set such that every truth function can be represented by a statement form containing only connectives from that set.

▷ One of the consequences of the preceding discussion is that $\{\sim, \wedge, \vee\}$ is an adequate set of connectives. We can use this to find other adequate sets.

Proposition 1.24

The pairs $\{\sim, \wedge\}$, $\{\sim, \vee\}$, and $\{\sim, \rightarrow\}$ are adequate sets of connectives.

Proof. First, for any statement forms $\mathcal{A}$ and $\mathcal{B}$, $(\mathcal{A} \vee \mathcal{B})$ is logically equivalent to $(\sim((\sim\mathcal{A}) \wedge (\sim\mathcal{B})))$, so any statement form containing just $\{\sim, \wedge, \vee\}$ can be transformed into a logically equivalent statement form containing only $\sim$ and $\wedge$.

Second, we can similarly use the logical equivalence of $(\mathcal{A} \wedge \mathcal{B})$ with $(\sim((\sim\mathcal{A}) \vee (\sim\mathcal{B})))$ to see that $\{\sim, \vee\}$ is adequate.

Third, we have to find statement forms logically equivalent to $(\mathcal{A} \wedge \mathcal{B})$ and $(\mathcal{A} \vee \mathcal{B})$ which involve only $\sim$ and $\rightarrow$.

$(\mathcal{A} \wedge \mathcal{B})$ is logically equivalent to $(\sim(\mathcal{A} \rightarrow (\sim\mathcal{B})))$,
$(\mathcal{A} \vee \mathcal{B})$ is logically equivalent to $((\sim\mathcal{A}) \rightarrow \mathcal{B})$.

These can be used to transform any statement form involving $\{\sim, \wedge, \vee\}$ into a logically equivalent statement form involving only $\sim$ and $\rightarrow$.

Example 1.25

$(((\sim p_1) \vee p_2) \rightarrow p_3)$ is logically equivalent to each of the following:

(a) $(\sim((\sim p_1) \vee p_2) \vee p_3)$.
(b) $\sim(\sim(p_1 \wedge (\sim p_2)) \wedge (\sim p_3))$.
(c) $((p_1 \rightarrow p_2) \rightarrow p_3)$.

▷ From our five connectives there are three ways of choosing an adequate pair. No other pair is adequate. To see this, consider first a pair of connectives neither of which is '~', and ask: can a truth function which always takes value F be expressed by a statement form using only such a pair of connectives? The answer must be in the negative, since giving all the statement variables in such a statement form the value T would necessarily give the whole statement form the value T. There is no way that it or any part of it can take value F under this assignment of truth values. So no statement form involving only connectives from $\wedge$, $\vee$, $\rightarrow$, $\leftrightarrow$ can be a contradiction. Thus no subset of this set of connectives can be adequate. The reader is left to verify that $\{\sim, \leftrightarrow\}$ is not an adequate set.

There are other connectives; indeed each truth table could be taken to define a connective, but their intuitive meanings would be less clear. However, there are two which deserve mention.

Nor

This is indicated by $\downarrow$, and the truth table is as follows:

p	q	$(p \downarrow q)$
T	T	F
T	F	F
F	T	F
F	F	T

Nand

This is denoted by $|$, and the truth table is as follows:

p	q	$p\|q$
T	T	F
T	F	T
F	T	T
F	F	T

The reason for interest in these connectives (and this has consequences in the design and study of computing machines) is given in the next proposition.

Proposition 1.26

The singleton sets $\{\downarrow\}$ and $\{|\}$ are adequate sets of connectives, i.e. Every truth function can be expressed by a statement form in which only $\downarrow$ (respectively $|$) appears.

Proof. We have only to express $\sim$ and $\wedge$, or $\sim$ and $\vee$ in terms of $\downarrow$

and in terms of $|$, since we know that $\{\sim, \wedge\}$ and $\{\sim, \vee\}$ are adequate sets. First,

$$(\sim p) \text{ is logically equivalent to } (p \downarrow p),$$

and

$$(p \wedge q) \text{ is logically equivalent to } ((p \downarrow p) \downarrow (q \downarrow q)).$$

Second,

$$(\sim p) \text{ is logically equivalent to } (p|p),$$

and

$$(p \vee q) \text{ is logically equivalent to } ((p|p)|(q|q)).$$

The verification is as usual by constructing the truth tables, and is left as an exercise.

Example 1.27

Find a statement form involving only $\downarrow$ which is logically equivalent to $(p \rightarrow q)$.

$$(p \rightarrow q) \text{ is logically equivalent to } \sim(p \wedge (\sim q))$$

and so to

$$\sim(p \wedge (q \downarrow q))$$

and so to

$$\sim((p \downarrow p) \downarrow [(q \downarrow q) \downarrow (q \downarrow q)])$$

and so to

$$\{(p \downarrow p) \downarrow [(q \downarrow q) \downarrow (q \downarrow q)]\} \downarrow \{(p \downarrow p) \downarrow [(q \downarrow q) \downarrow (q \downarrow q)]\}.$$

This example illustrates what one has to pay in terms of complication and length if one wishes to use only the one connective.

Exercises

14   Find statement forms in which only the connectives $\sim$ and $\vee$ occur, which are logically equivalent to the following:
   (a) $(p \rightarrow (q \rightarrow r))$;
   (b) $(((\sim p) \wedge (\sim q)) \rightarrow ((\sim r) \wedge s))$;
   (c) $(p \leftrightarrow q)$.
15   Find statement forms in which only the connectives $\sim$ and $\wedge$ occur, which are logically equivalent to the following:
   (a) $(p \rightarrow (q \rightarrow r))$;

    (b) $((p \lor q \lor r) \land ((\sim p) \lor (\sim q) \lor (\sim r)))$;

    (c) $((p \leftrightarrow (\sim q)) \leftrightarrow r)$.

16   Find statement forms in which only the connectives $\sim$ and $\rightarrow$ occur, which are logically equivalent to the following:

    (a) $((p \land q) \lor (r \land s))$;

    (b) $(p \leftrightarrow q)$;

    (c) $(p \land q \land r)$.

17   (a) Prove that $\{\land, \lor\}$ is not an adequate set of connectives.

    (b) (Harder) Prove that $\{\sim, \leftrightarrow\}$ is not an adequate set of connectives.

18   Find a statement in which only the connective $|$ occurs, which is logically equivalent to $(p \rightarrow q)$.

19   Prove that there is no binary connective other than $\downarrow$ and $|$ which itself constitutes an adequate set of connectives. (Hint: consider the truth table of any such connective.)

1.6 Arguments and validity

Let us now return to consideration of arguments. For the moment we are necessarily restricted to arguments whose premisses and conclusion are all simple or compound statements in the sense defined at the beginning of the chapter. We saw that what was important was the 'form' of the argument rather than the meanings of the statements involved. We shall consider, therefore, *argument forms*. In a previous example we came across the argument form

$$(p \rightarrow q)$$

$$p$$

$$\therefore \quad q$$

In general, an argument form is a finite sequence of statement forms, the last of which is regarded as the conclusion and the remainder as premisses.

In deciding and defining what constitutes a 'valid' argument form we come up against the same kind of difficulty that we had over the implication symbol. When we assign truth values to the statement variables appearing in an argument form we may find that the conclusion is false and one or more of the premisses is also false. Do false premisses justify a false conclusion? In a sense it is an irrelevant question, for in normal usage we use an argument only in order to demonstrate that a certain conclusion follows from known premisses. All we shall require of a valid argument form, therefore, is that, under any assignment of truth values to the statement variables, if the premisses all take value T then the conclusion also takes value T. Equivalently, we can make the following definition.

Definition 1.28

The argument form

$$\mathscr{A}_1, \mathscr{A}_2, \ldots, \mathscr{A}_n; \quad \therefore \mathscr{A}$$

is *invalid* if it is possible to assign truth values to the statement variables occurring in such a way as to make each of $\mathscr{A}_1, \ldots, \mathscr{A}_n$ take value T and to make $\mathscr{A}$ take value F. Otherwise the argument form is *valid*.

▷ We now have the problem of testing a given argument form for validity. Let us consider our simple example: $(p \to q), p; \therefore q$. Construct a truth table for all the statement forms appearing as premisses or conclusion.

p	q	$(p \to q)$
T	T	T
T	F	F
F	T	T
F	F	T

It is only in the first row that the premisses both take value T, and it so happens that the conclusion also takes value T in this row. So the argument form is not invalid, i.e. it is valid.

Example 1.29

Investigate the validity of the argument form:

$$(p \to q), ((\sim q) \to r), r; \therefore p.$$

Construct a truth table:

$(p$	$\to$	$q)$	$((\sim$	$q)$	$\to$	$r)$	r	p	
T	T	T	F	T	T	T	T	T	←
T	T	T	F	T	T	F	F	T	
T	F	F	T	F	T	T	T	T	
T	F	F	T	F	F	F	F	T	
F	T	T	F	T	T	T	T	F	←
F	T	T	F	T	T	F	F	F	
F	T	F	T	F	T	T	T	F	←
F	T	F	T	F	T	F	F	F	

The three arrowed rows are those in which all the premisses take value T. In the fifth and seventh rows, however, the conclusion takes value F. Hence the argument form is invalid.

▷ So here is a method for deciding the validity of an argument form which will give us the answer in every case. However, if the number of different statement variables is large, our truth table will be unwieldy and impractical. In any case we do not need the whole truth table for our purpose. Our method is a search for a row of a particular kind, and we can conduct such a search in a systematic way, rather than in the trial and error way of constructing the whole truth table. The practical procedure is best described by means of an example.

Example 1.30

Test the validity of the following argument form.

$$((\sim p_1) \vee p_2), (p_1 \to (p_3 \wedge p_4)), (p_4 \to p_2); \therefore (p_2 \vee p_3).$$

We attempt to assign truth values so as to demonstrate the invalidity of the argument form, i.e. so as to make the premisses true and the conclusion false. To make $(p_2 \vee p_3)$ take value F we must assign F to each of p_2 and p_3. Then to make $(p_4 \to p_2)$ take value T we must assign F to p_4. To make $((\sim p_1) \vee p_2)$ take value T we must assign F to p_1. Now check the truth value of the other premisses $(p_1 \to (p_3 \wedge p_4))$ under these assignments. It is T. Hence

p_1	p_2	p_3	p_4
F	F	F	F

is an assignment of truth values under which the premisses all take value T and the conclusion takes value F. The argument form is therefore invalid.

Notice that if the argument form were valid, we would be unable to assign truth values as we set out to do.

Example 1.31

Test the validity of the argument

$$(p_1 \to (p_2 \to p_3)), p_2; \therefore (p_1 \to p_3).$$

Attempt to assign truth values to demonstrate the invalidity of the argument form. To make $(p_1 \to p_3)$ take value F we require p_1 to have value T and p_3 to have value F. We also require p_2 to be T. Under this assignment of truth values, the other premiss, $(p_1 \to (p_2 \to p_3))$ takes value F. So it is impossible to assign values so that the premisses are true and the conclusion is false, and this argument is valid.

▷ The next proposition makes explicit the connection between arguments and implications which are mentioned briefly above.

Proposition 1.32

The argument form

$$\mathcal{A}_1, \ldots, \mathcal{A}_n; \therefore \mathcal{A}$$

is valid if and only if the statement form

$$((\mathcal{A}_1 \wedge \ldots \wedge \mathcal{A}_n) \to \mathcal{A})$$

is a tautology.

Proof. Suppose first that $\mathcal{A}_1, \ldots, \mathcal{A}_n; \therefore \mathcal{A}$ is a valid argument form and that $((\mathcal{A}_1 \wedge \ldots \wedge \mathcal{A}_n) \to \mathcal{A})$ is *not* a tautology. Then there is an assignment of truth values to the statement variables occurring such that $(\mathcal{A}_1 \wedge \ldots \wedge \mathcal{A}_n)$ takes value T, and $\mathcal{A}$ takes value F. For this assignment, then, each $\mathcal{A}_i$ takes value T $(1 \le i \le n)$, and $\mathcal{A}$ takes value F. This contradicts the validity of the argument form, and so $((\mathcal{A}_1 \wedge \ldots \wedge \mathcal{A}_n) \to \mathcal{A})$ must be a tautology.

Now suppose, conversely, that $((\mathcal{A}_1 \wedge \ldots \wedge \mathcal{A}_n) \to \mathcal{A})$ is a tautology and that $\mathcal{A}_1, \ldots, \mathcal{A}_n; \therefore \mathcal{A}$ is *not* a valid argument form. Then there is an assignment of truth values which makes each $\mathcal{A}_i$ $(1 \le i \le n)$ take value T and makes $\mathcal{A}$ take value F, so that $((\mathcal{A}_1 \wedge \ldots \wedge \mathcal{A}_n) \to \mathcal{A})$ takes value F. This contradicts the assumption that this statement form is a tautology, and so $\mathcal{A}_1, \ldots, \mathcal{A}_n; \therefore \mathcal{A}$ is a valid argument form.

▷ Let us conclude the chapter with a remark about the familiar mathematical method of 'proof by contradiction' or *reductio ad absurdum*, which was used, incidentally, in the proof above.

Such a proof consists of a deduction of a contradiction from the negation of the statement whose proof is required. That this is a legitimate procedure in the terms of this chapter can be seen as follows. If we have an argument which is known to be an instance of a valid argument form, and its conclusion is known to be false, then at least one of the premisses must be false. If all the premisses are known to be true except one (the assumed one), then the legitimate deduction is that this assumed one is the one which is false.

Exercises

20 For each of the following arguments, write down a corresponding argument form and determine whether it is valid or invalid.

   (*a*) If the function f is not continuous then the function g is not differentiable. g is differentiable. Therefore f is continuous.

(b) If Smith has installed central heating, then either he has sold his car or he has borrowed money from the bank. Smith has not borrowed money from the bank. Therefore, if Smith has not sold his car, then he has not installed central heating.

(c) If there is oil in Polygonia then either the experts are right or the government is lying. There is no oil in Polygonia, or else the experts are wrong. Therefore the government is not lying.

(d) If U is a subspace of V then U is a subset of V, U contains the zero vector and U is closed. U is a subset of V, and if U is closed then U contains the zero vector. Therefore, if U is closed then U is a subspace of V.

21 Suppose that $\mathscr{A}_1, \mathscr{A}_2, \ldots, \mathscr{A}_n; \therefore \mathscr{A}$ is a valid argument form. Prove that $\mathscr{A}_1, \mathscr{A}_2, \ldots, \mathscr{A}_{n-1}; \therefore (\mathscr{A}_n \to \mathscr{A})$ is also a valid argument form.

22 Show that the following is a valid argument form:

$$p, (p|(q|r)); \therefore r.$$

2
Formal statement calculus

2.1 The formal system *L*

Our study of logic is aimed, in part at least, at an analysis of the process of deduction. In the first chapter we saw how to abstract the forms of statements and arguments in order to see more clearly the relationships between them and to give an intuitive definition of a valid argument. Certain questions remain, however. For example, can we find a simple procedure which enables us to construct an argument step by step, in the knowledge that at every step our argument is valid? On what could we base such a procedure? For we cannot deduce from nothing, and we must make some original assumptions. To investigate such questions we introduce the idea of a *formal deductive system*. This is in essence a continuation of our procedure of abstraction, in which we make abstract the notion of proof. The word 'formal' is one which appears regularly in logic textbooks without being explained. It is used when referring to a situation where symbols are being used and where the behaviour and properties of the symbols are determined completely by a given set of rules. In a formal system the symbols have no meanings, and in dealing with them we must be careful to assume nothing about their properties other than what is specified in the system. It is only in this way that we can be sure that all the assumptions that we make when we follow through a deduction are explicit, and it is only by making all our assumptions explicit that we can discover anything fundamental about logic.

In this book we shall be concerned with two particular formal systems, but sometimes we shall require to deal with others which are modifications of these two, so let us start by giving a general definition of what constitutes a formal system.

To specify a formal system we require:

1. An alphabet of symbols.

2. A set of finite strings of these symbols, called well-formed formulas. These are to be thought of as the words and sentences in our formal languages.

3. A set of well-formed formulas, called axioms.

4. A finite set of 'rules of deduction', i.e., rules which enable one to deduce a well-formed formula $\mathcal{A}$, say, as a 'direct consequence' of a finite set of well-formed formulas $\mathcal{A}_1, \ldots, \mathcal{A}_k$, say.

Given these four things we can follow through deductive procedures (which may or may not bear any relation to logical deduction, depending on the particular formal system under consideration) by successive application of the rules of deduction starting with axioms. But we shall make this precise shortly.

Notation. We shall abbreviate 'well-formed formula' by '*wf.*' from now on.

Definition 2.1

The *formal system L of statement calculus* is defined by the following:

1. Alphabet of symbols (infinite):

$$\sim, \rightarrow, (\ , \ ), p_1, p_2, p_3, \ldots$$

2. Set of *wfs*. Instead of specifying the set explicitly we give an inductive rule in three parts (see Definition 1.2):

(i) p_i is a *wf.* for each $i \geq 1$.

(ii) If $\mathcal{A}$ and $\mathcal{B}$ are *wfs.* then $(\sim\mathcal{A})$ and $(\mathcal{A} \rightarrow \mathcal{B})$ are *wfs.*

(iii) The set of all *wfs.* is generated by (i) and (ii).

3. Axioms. There are infinitely many axioms, so we cannot list them all. However we can specify all of them by means of three *axiom schemes*. For any *wfs.* $\mathcal{A}$, $\mathcal{B}$, $\mathcal{C}$, the following *wfs.* are axioms of L:

(L1) $(\mathcal{A} \rightarrow (\mathcal{B} \rightarrow \mathcal{A}))$.

(L2) $((\mathcal{A} \rightarrow (\mathcal{B} \rightarrow \mathcal{C})) \rightarrow ((\mathcal{A} \rightarrow \mathcal{B}) \rightarrow (\mathcal{A} \rightarrow \mathcal{C})))$.

(L3) $(((\sim\mathcal{A}) \rightarrow (\sim\mathcal{B})) \rightarrow (\mathcal{B} \rightarrow \mathcal{A}))$.

Note that each axiom scheme has infinitely many 'instances', as $\mathcal{A}$, $\mathcal{B}$, $\mathcal{C}$ range over all *wfs.* of L.

4. Rules of deduction. In L there is only one rule of deduction, namely *modus ponens* (abbreviated by *MP*). It says: from $\mathcal{A}$ and $(\mathcal{A} \rightarrow \mathcal{B})$, $\mathcal{B}$ is a direct consequence, where $\mathcal{A}$ and $\mathcal{B}$ are any *wfs.* of L.

▷ The alphabet and the set of *wfs.* have been chosen to reflect the development of the previous chapter. We wish the *wfs.* to stand in some way for statement forms, so the definition corresponds closely to the definition of statement form. The symbols $\wedge$, $\vee$, $\leftrightarrow$ do not appear in the alphabet for L, so expressions in which they occur are not part of L. However, as we have seen, $\{\sim, \rightarrow\}$ is an adequate set of connectives, so every truth function will be represented by some *wf.* of L, and every statement form will be logically equivalent to some *wf.* of L. (Be careful to note, however, that statement forms and logical equivalence are notions from Chapter 1 and have no place in the formal system L.) In L we have limited the number of connectives in order to make the formal language simpler and in order that a set of axioms and rules of deduction can be written down compactly. If we had included $\wedge$, say, in the alphabet of symbols, we would also have had to include axioms and/or rules of deduction to govern its behaviour (and to make explicit its connection with the symbol $\rightarrow$), for the symbols in our language have no presupposed properties. All their properties must be derivable from the information in the definition of L.

The rule of deduction in L seems intuitively to be a reasonable one. This rule corresponds to one of the standard ways of proceeding in an argument in everyday language. The axioms of L are the least obvious part of the system. The reader should examine them closely and notice that, if they are regarded as statement forms, they are all tautologies. The axioms must be there to give a basis that we can deduce from, and there is nothing unique about the set we have chosen. The above axiom schemes happen to be convenient for the proofs of two later pro-positions, the Deduction Theorem and the Adequacy Theorem. In the course of the exposition, the reasons for the choice of axioms should become clearer.

We must now explain the deductive nature of L.

Definition 2.2

A *proof* in L is a sequence of *wfs.* $\mathscr{A}_1, \ldots, \mathscr{A}_n$ such that for each i $(1 \le i \le n)$, either $\mathscr{A}_i$ is an axiom of L or $\mathscr{A}_i$ follows from two previous members of the sequence, say $\mathscr{A}_j$ and $\mathscr{A}_k$ $(j < i, k < i)$ as a direct consequence using the rule of deduction *MP*. Such a proof will be referred to as a *proof of* $\mathscr{A}_n$ in L, and $\mathscr{A}_n$ is said to be a *theorem of* L.

Remarks 2.3

(*a*) In the above definition, observe that $\mathscr{A}_j$ and $\mathscr{A}_k$ must necessarily be of the forms $\mathscr{B}$ and $(\mathscr{B} \rightarrow \mathscr{A}_i)$, or vice versa.

(*b*) If $\mathscr{A}_1, \ldots, \mathscr{A}_n$ is a proof in L and $k < n$ then $\mathscr{A}_1, \ldots, \mathscr{A}_k$ is also a proof in L (it clearly satisfies the definition), and so $\mathscr{A}_k$ is a theorem of L.

(*c*) Axioms of L are certainly theorems of L. Their proofs in L are one-member sequences.

Example 2.4

The following sequence is a proof in L.

(1) $(p_1 \rightarrow (p_2 \rightarrow p_1))$ (instance of $(L1)$)

(2) $((p_1 \rightarrow (p_2 \rightarrow p_1)) \rightarrow ((p_1 \rightarrow p_2) \rightarrow (p_1 \rightarrow p_1)))$ ($L2$)

(3) $((p_1 \rightarrow p_2) \rightarrow (p_1 \rightarrow p_1))$ (from (1), (2) by MP).

It follows that $((p_1 \rightarrow p_2) \rightarrow (p_1 \rightarrow p_1))$ is a theorem of L.

▷ A proof in L is a deduction starting from the axioms. We shall need also the more general notion of deduction from some given set of *wfs*.

Definition 2.5

Let Γ be a set of *wfs*. of L (which may or may not be axioms or theorems of L). A sequence $\mathcal{A}_1, \ldots, \mathcal{A}_n$ of *wfs*. of L is a *deduction from Γ* if for each i $(1 \leq i \leq n)$, one of the following holds:
  (a) $\mathcal{A}_i$ is an axiom of L,
  (b) $\mathcal{A}_i$ is a member of Γ, or
  (c) $\mathcal{A}_i$ follows from two previous members of the sequence as a direct consequence using MP.
    So a deduction from Γ is just a 'proof' in which the members of Γ are considered as temporary axioms.
    The last member, $\mathcal{A}_n$, of a sequence which is a deduction from Γ is said to be *deducible from Γ*, or to be a consequence of Γ in L.
    If a *wf*. $\mathcal{A}$ is the last member of some deduction from Γ, we say Γ *yields* $\mathcal{A}$ and write $\Gamma \vdash_L \mathcal{A}$.
    Notice that a theorem of L is deducible from the empty set of *wfs*. (a proof in L is a deduction from $\varnothing$), so if $\mathcal{A}$ is a theorem of L we may write $\varnothing \vdash_L \mathcal{A}$, or more normally for brevity, just $\vdash_L \mathcal{A}$.

Remark. It is important to remember that '$\vdash$' is not a symbol of L, and so any expression in which it appears cannot be a part of L. For example, $\vdash_L \mathcal{A}$, far from being part of L, is a statement *about* L, namely the statement that the *wf*. $\mathcal{A}$ is a theorem of L.

Example 2.6

The following is a deduction in L which establishes $\{\mathcal{A}, (\mathcal{B} \rightarrow (\mathcal{A} \rightarrow \mathcal{C}))\} \vdash_L (\mathcal{B} \rightarrow \mathcal{C})$, where $\mathcal{A}, \mathcal{B}, \mathcal{C}$ stand for any *wfs*. of L.

(1)	$\mathscr{A}$	assumption
(2)	$(\mathscr{B} \to (\mathscr{A} \to \mathscr{C}))$	assumption
(3)	$(\mathscr{A} \to (\mathscr{B} \to \mathscr{A}))$	(L1)
(4)	$(\mathscr{B} \to \mathscr{A})$	(1), (3) MP
(5)	$((\mathscr{B} \to (\mathscr{A} \to \mathscr{C})) \to ((\mathscr{B} \to \mathscr{A}) \to (\mathscr{B} \to \mathscr{C})))$	(L2)
(6)	$((\mathscr{B} \to \mathscr{A}) \to (\mathscr{B} \to \mathscr{C}))$	(2), (5) MP
(7)	$(\mathscr{B} \to \mathscr{C})$	(4), (6) MP.

▷ This example and the remark preceding it bring to light a distinction which it is important to emphasise. Just as $\vdash_L \mathscr{A}$ is a statement about L, so the result of the example is a general result about L:
For any wfs. $\mathscr{A}$, $\mathscr{B}$, $\mathscr{C}$ of L,

$$\{\mathscr{A}, (\mathscr{B} \to (\mathscr{A} \to \mathscr{C}))\} \vdash_L (\mathscr{B} \to \mathscr{C}).$$

This result about L is certainly not part of L. There are two levels to our proceedings – we are proving results about proofs. We shall use the word 'theorem' only to refer to wfs. in formal systems which have proofs in the sense of Definition 2.2. The word 'metatheorem' is sometimes used for results, such as the above mentioned, which are results about formal systems. Theorems are wfs. of a particular kind, metatheorems are written in ordinary mathematical language. Our use of the word proposition is thus to avoid confusion. Generally speaking, our propositions are metatheorems.

Note also that the script letters $\mathscr{A}$, $\mathscr{B}$, etc. which we have been using are not part of L. We use them as a convenience, to stand for unspecified wfs. of L, or when we are making general assertions about L.

We have set up the formal system L as a system in which certain wfs. can be demonstrated to be theorems. Naturally we are interested in which wfs. of L are theorems. Now the only method we have for showing that a wf. is a theorem is actually writing out a sequence of wfs. which constitutes a proof. This can be a very lengthy business, and one in which the methods to be used in particular cases are not always obvious.

Example 2.7

For any wfs. $\mathscr{A}$ and $\mathscr{B}$ of L

(a) $\vdash_L (\mathscr{A} \to \mathscr{A})$,

(b) $\vdash_L (\sim\mathscr{B} \to (\mathscr{B} \to \mathscr{A}))$.

We write out a proof in L in each case. For (a):

(1) $(\mathscr{A} \to ((\mathscr{A} \to \mathscr{A}) \to \mathscr{A})) \to ((\mathscr{A} \to (\mathscr{A} \to \mathscr{A})) \to (\mathscr{A} \to \mathscr{A}))$ $(L2)$
(2) $(\mathscr{A} \to ((\mathscr{A} \to \mathscr{A}) \to \mathscr{A}))$ $(L1)$
(3) $((\mathscr{A} \to (\mathscr{A} \to \mathscr{A})) \to (\mathscr{A} \to \mathscr{A}))$ $(1), (2)\, MP$
(4) $(\mathscr{A} \to (\mathscr{A} \to \mathscr{A}))$ $(L1)$
(5) $(\mathscr{A} \to \mathscr{A})$ $(3), (4)\, MP.$

For (b):

(1) $(\sim\!\mathscr{B} \to (\sim\!\mathscr{A} \to \sim\!\mathscr{B}))$ $(L1)$
(2) $((\sim\!\mathscr{A} \to \sim\!\mathscr{B}) \to (\mathscr{B} \to \mathscr{A}))$ $(L3)$
(3) $((\sim\!\mathscr{A} \to \sim\!\mathscr{B}) \to (\mathscr{B} \to \mathscr{A})) \to (\sim\!\mathscr{B} \to ((\sim\!\mathscr{A} \to \sim\!\mathscr{B})$
   $\to (\mathscr{B} \to \mathscr{A})))$ $(L1)$
(4) $(\sim\!\mathscr{B} \to ((\sim\!\mathscr{A} \to \sim\!\mathscr{B}) \to (\mathscr{B} \to \mathscr{A})))$ $(2), (3)\, MP$
(5) $(\sim\!\mathscr{B} \to ((\sim\!\mathscr{A} \to \sim\!\mathscr{B}) \to (\mathscr{B} \to \mathscr{A}))) \to ((\sim\!\mathscr{B} \to (\sim\!\mathscr{A} \to \sim\!\mathscr{B})$
   $\to (\sim\!\mathscr{B} \to (\mathscr{B} \to \mathscr{A})))$ $(L2)$
(6) $(\sim\!\mathscr{B} \to (\sim\!\mathscr{A} \to \sim\!\mathscr{B})) \to (\sim\!\mathscr{B} \to (\mathscr{B} \to \mathscr{A}))$ $(4), (5)\, MP$
(7) $(\sim\!\mathscr{B} \to (\mathscr{B} \to \mathscr{A}))$ $(1), (6)\, MP.$

In the above we have been less strict in our use of parentheses than previously. Some of the expressions used in the proof for (b) are not *wfs*. For example, (1) should be $((\sim\!\mathscr{B}) \to ((\sim\!\mathscr{A}) \to (\sim\!\mathscr{B})))$. However, we shall continue to abbreviate *wfs*. in this way. The advantages to be gained are obvious, but we must take care not to omit so many parentheses that the resulting expression is ambiguous.

▷ One way in which theorem proving can be made less arduous is to allow in a proof the insertion of a *wf*. for which a proof in L has previously been obtained. This corresponds to the standard mathematical procedure of quoting previously proved theorems. Another way is to make use of some general metatheorems, some of which have the effect of additional rules of inference. The main tool is the following.

Proposition 2.8 (The Deduction Theorem)

If $\Gamma \cup \{\mathscr{A}\} \underset{L}{\vdash} \mathscr{B}$ then $\Gamma \underset{L}{\vdash} (\mathscr{A} \to \mathscr{B})$, where $\mathscr{A}$ and $\mathscr{B}$ are *wfs*. of L, and Γ is a set of *wfs*. of L (possibly empty).

Proof. The proof is by induction on the number of *wfs*. in the sequence forming the deduction of $\mathscr{B}$ from $\Gamma \cup \{\mathscr{A}\}$. For the base step suppose that this sequence has one member. That member must be $\mathscr{B}$ itself, and so either $\mathscr{B}$ is an axiom of L or $\mathscr{B}$ is a member of $\Gamma \cup \{\mathscr{A}\}$.

Case 1: $\mathscr{B}$ is an axiom of L. The following is then a deduction of $(\mathscr{A} \to \mathscr{B})$ from Γ.

(1) $\mathscr{B}$ axiom of L
(2) $(\mathscr{B} \to (\mathscr{A} \to \mathscr{B}))$ $(L1)$
(3) $(\mathscr{A} \to \mathscr{B})$ $(1), (2)\, MP.$

Hence $\Gamma \vdash_L (\mathscr{A} \to \mathscr{B})$.

  Case 2: $\mathscr{B} \in \Gamma$. The following deduction shows that $\Gamma \vdash_L (\mathscr{A} \to \mathscr{B})$.

(1) $\mathscr{B}$ member of Γ
(2) $(\mathscr{B} \to (\mathscr{A} \to \mathscr{B}))$ (L1)
(3) $(\mathscr{A} \to \mathscr{B})$ (1), (2) *MP*.

  Case 3: $\mathscr{B}$ is $\mathscr{A}$. We have seen that $\vdash_L (\mathscr{A} \to \mathscr{A})$, so the proof in *L* of $(\mathscr{A} \to \mathscr{A})$ will serve as a deduction of $(\mathscr{A} \to \mathscr{A})$ from Γ. Hence in this case also we have $\Gamma \vdash_L (\mathscr{A} \to \mathscr{B})$. This ends the base step.

  Now suppose that the deduction of $\mathscr{B}$ from $\Gamma \cup \{\mathscr{A}\}$ is a sequence with *n* members, where $n > 1$, and that the proposition holds for all *wfs.* $\mathscr{C}$ which can be deduced from $\Gamma \cup \{\mathscr{A}\}$ via a sequence with fewer than *n* members. This time there are four cases to consider.

Case 1: $\mathscr{B}$ is an axiom of *L*. Precisely as in Case 1 above, we show that $\Gamma \vdash_L (\mathscr{A} \to \mathscr{B})$.

Case 2: $\mathscr{B} \in \Gamma$. Again, $\Gamma \vdash_L (\mathscr{A} \to \mathscr{B})$ just as in Case 2 above.

Case 3: $\mathscr{B}$ is $\mathscr{A}$. As Case 3 above.

Case 4: $\mathscr{B}$ is obtained from two previous *wfs.* in the deduction by an application of *MP*. These two *wfs.* must have the forms $\mathscr{C}$ and $(\mathscr{C} \to \mathscr{B})$, and each one can certainly be deduced from $\Gamma \cup \{\mathscr{A}\}$ by a sequence with fewer than *n* members. In each case just omit the subsequent members from the original deduction and what remains is the desired sequence (cf. Remark 2.3(*b*)). We have $\Gamma \cup \{\mathscr{A}\} \vdash_L \mathscr{C}$ and $\Gamma \cup \{\mathscr{A}\} \vdash_L (\mathscr{C} \to \mathscr{B})$, and, applying the hypothesis of induction,

$$\Gamma \vdash_L (\mathscr{A} \to \mathscr{C}) \quad \text{and} \quad \Gamma \vdash_L (\mathscr{A} \to (\mathscr{C} \to \mathscr{B})).$$

The required deduction of $(\mathscr{A} \to \mathscr{B})$ from Γ may now be given as follows:

(1)
 $\vdots$ $\left.\rule{0pt}{32pt}\right\}$ deduction of $(\mathscr{A} \to \mathscr{C})$ from Γ
(*k*) $(\mathscr{A} \to \mathscr{C})$

(*k* + 1)
 $\vdots$ $\left.\rule{0pt}{32pt}\right\}$ deduction of $(\mathscr{A} \to (\mathscr{C} \to \mathscr{B}))$ from Γ
(*l*) $(\mathscr{A} \to (\mathscr{C} \to \mathscr{B}))$

(*l* + 1) $(\mathscr{A} \to (\mathscr{C} \to \mathscr{B})) \to ((\mathscr{A} \to \mathscr{C}) \to (\mathscr{A} \to \mathscr{B}))$ (L2)

(*l* + 2) $(\mathscr{A} \to \mathscr{C}) \to (\mathscr{A} \to \mathscr{B})$ (*l*), (*l* + 1) *MP*

$(l+3)$ $(\mathscr{A} \to \mathscr{B})$ $(k), (l+2)\, MP.$

$\therefore$ $\Gamma \underset{L}{\vdash} (\mathscr{A} \to \mathscr{B})$ in all four cases.

Hence by the principle of mathematical induction the proposition holds whatever the number of wfs. in the deduction of $\mathscr{B}$ from $\Gamma \cup \{\mathscr{A}\}$.

Note. We have not used any instance of ($L3$) in the proof of this proposition. This has important consequences in the study of other formal systems which have different sets of axioms.

▷ The converse of the Deduction Theorem is easier to prove.

Proposition 2.9

If $\Gamma \underset{L}{\vdash} (\mathscr{A} \to \mathscr{B})$ then $\Gamma \cup \{\mathscr{A}\} \underset{L}{\vdash} \mathscr{B}$, where $\mathscr{A}$ and $\mathscr{B}$ are wfs. of L and Γ is a (possibly empty) set of wfs. of L.

Proof. Given a deduction of $(\mathscr{A} \to \mathscr{B})$ from Γ, we wish to construct a deduction of $\mathscr{B}$ from $\Gamma \cup \{\mathscr{A}\}$.

$$
\left.
\begin{array}{ll}
(1) & \\
\quad \vdots & \\
(k) & (\mathscr{A} \to \mathscr{B})
\end{array}
\right\} \text{deduction of } (\mathscr{A} \to \mathscr{B}) \text{ from } \Gamma
$$

$(k+1)$ $\mathscr{A}$ member of $\Gamma \cup \{\mathscr{A}\}$

$(k+2)$ $\mathscr{B}$ $(k), (k+1)\, MP.$

▷ The use of the Deduction Theorem is illustrated in the proof of the following, which may be used as a new rule of deduction.

Corollary 2.10

For any wfs. $\mathscr{A}, \mathscr{B}, \mathscr{C}$ of L,

$$\{(\mathscr{A} \to \mathscr{B}), (\mathscr{B} \to \mathscr{C})\} \underset{L}{\vdash} (\mathscr{A} \to \mathscr{C}).$$

Proof. We write out the deduction.

(1)	$(\mathscr{A} \to \mathscr{B})$	assumption
(2)	$(\mathscr{B} \to \mathscr{C})$	assumption
(3)	$\mathscr{A}$	assumption
(4)	$\mathscr{B}$	$(1), (3)\, MP$
(5)	$\mathscr{C}$	$(2), (4)\, MP.$

What we have demonstrated is

$$\{(\mathcal{A} \to \mathcal{B}), (\mathcal{B} \to \mathcal{C}), \mathcal{A}\} \vdash_{L} \mathcal{C} \qquad\qquad (*)$$

i.e.

$$\{(\mathcal{A} \to \mathcal{B}), (\mathcal{B} \to \mathcal{C})\} \cup \{\mathcal{A}\} \vdash_{L} \mathcal{C}$$

So by the Deduction Theorem, we have

$$\{(\mathcal{A} \to \mathcal{B}), (\mathcal{B} \to \mathcal{C})\} \vdash_{L} (\mathcal{A} \to \mathcal{C}) \text{ as required.}$$

▷ This result will be applied several times in what follows, and we shall refer to it as the rule '*hypothetical syllogism*', and abbreviate it *HS*.

Note. There are several ways of applying the Deduction Theorem to $(*)$ above. We can also deduce each of the following:

$$\{(\mathcal{A} \to \mathcal{B}), \mathcal{A}\} \vdash_{L} ((\mathcal{B} \to \mathcal{C}) \to \mathcal{C}),$$

$$\{(\mathcal{B} \to \mathcal{C}), \mathcal{A}\} \vdash_{L} ((\mathcal{A} \to \mathcal{B}) \to \mathcal{C}).$$

By applying the Deduction Theorem again to the result of the corollary we obtain

$$\{(\mathcal{A} \to \mathcal{B})\} \vdash_{L} ((\mathcal{B} \to \mathcal{C}) \to (\mathcal{A} \to \mathcal{C})),$$

and hence

$$\vdash_{L} ((\mathcal{A} \to \mathcal{B}) \to ((\mathcal{B} \to \mathcal{C}) \to (\mathcal{A} \to \mathcal{C}))).$$

Proposition 2.11

For any *wfs.* $\mathcal{A}$ and $\mathcal{B}$ of *L*, the following are theorems of *L*.
    (*a*) $(\sim\mathcal{B} \to (\mathcal{B} \to \mathcal{A})),$
    (*b*) $((\sim\mathcal{A} \to \mathcal{A}) \to \mathcal{A}).$

Proof. ((a) appeared in example 2.7, but we include it here to illustrate the simplification produced by use of the new rule *HS*.)
For (a):
    (1) $(\sim\mathcal{B} \to (\sim\mathcal{A} \to \sim\mathcal{B}))$ (*L*1)
    (2) $(\sim\mathcal{A} \to \sim\mathcal{B}) \to (\mathcal{B} \to \mathcal{A})$ (*L*3)
    (3) $(\sim\mathcal{B} \to (\mathcal{B} \to \mathcal{A}))$ (1), (2) *HS*.
For (b):
    (1) $(\sim\mathcal{A} \to \mathcal{A})$ assumption

$$(2) \quad (\sim\!\mathcal{A} \to (\sim\sim(\sim\!\mathcal{A} \to \mathcal{A}) \to \sim\!\mathcal{A})) \qquad\qquad (L1)$$
$$(3) \quad (\sim\sim(\sim\!\mathcal{A} \to \mathcal{A}) \to \sim\!\mathcal{A}) \to (\mathcal{A} \to \sim(\sim\!\mathcal{A} \to \mathcal{A})) \qquad (L3)$$
$$(4) \quad (\sim\!\mathcal{A} \to (\mathcal{A} \to \sim(\sim\!\mathcal{A} \to \mathcal{A}))) \qquad\qquad (2),(3)\ HS$$
$$(5) \quad (\sim\!\mathcal{A} \to (\mathcal{A} \to \sim(\sim\!\mathcal{A} \to \mathcal{A}))) \to ((\sim\!\mathcal{A} \to \mathcal{A}) \to$$
$$(\sim\!\mathcal{A} \to \sim(\sim\!\mathcal{A} \to \mathcal{A}))) \qquad\qquad (L2)$$
$$(6) \quad (\sim\!\mathcal{A} \to \mathcal{A}) \to (\sim\!\mathcal{A} \to \sim(\sim\!\mathcal{A} \to \mathcal{A})) \qquad (4),(5)\ MP$$
$$(7) \quad (\sim\!\mathcal{A} \to \sim(\sim\!\mathcal{A} \to \mathcal{A})) \qquad\qquad (1),(6)\ MP$$
$$(8) \quad (\sim\!\mathcal{A} \to \sim(\sim\!\mathcal{A} \to \mathcal{A})) \to ((\sim\!\mathcal{A} \to \mathcal{A}) \to \mathcal{A}) \qquad (L3)$$
$$(9) \quad (\sim\!\mathcal{A} \to \mathcal{A}) \to \mathcal{A} \qquad\qquad (7),(8)\ MP$$
$$(10) \quad \mathcal{A} \qquad\qquad (1),(9)\ MP.$$

Hence $(\sim\!\mathcal{A} \to \mathcal{A}) \underset{L}{\vdash} \mathcal{A}$,

$\therefore \underset{L}{\vdash} ((\sim\!\mathcal{A} \to \mathcal{A}) \to \mathcal{A})$, by the Deduction Theorem. These results will be useful later.

Exercises

1 Write out proofs in L for the following *wfs*.
   - (a) $(p_1 \to p_2) \to ((\sim\!p_1 \to \sim\!p_2) \to (p_2 \to p_1))$;
   - (b) $((p_1 \to (p_2 \to p_3)) \to (p_1 \to p_2)) \to ((p_1 \to (p_2 \to p_3)) \to (p_1 \to p_3))$;
   - (c) $(p_1 \to (p_1 \to p_2)) \to (p_1 \to p_2)$;
   - (d) $(p_1 \to (p_2 \to (p_1 \to p_2)))$.

2 Show that the following hold for any *wfs*. $\mathcal{A}$, $\mathcal{B}$, $\mathcal{C}$ of L.
   - (a) $\{(\sim\!\mathcal{A})\} \underset{L}{\vdash} (\mathcal{A} \to \mathcal{B})$;
   - (b) $\{(\sim(\sim\!\mathcal{A}))\} \underset{L}{\vdash} \mathcal{A}$;
   - (c) $\{(\mathcal{A} \to \mathcal{B}), (\sim(\mathcal{B} \to \mathcal{C}) \to (\sim\!\mathcal{A}))\} \underset{L}{\vdash} (\mathcal{A} \to \mathcal{C})$;
   - (d) $\{(\mathcal{A} \to (\mathcal{B} \to \mathcal{C}))\} \underset{L}{\vdash} (\mathcal{B} \to (\mathcal{A} \to \mathcal{C}))$.

3 Using the Deduction Theorem for L, show that the following *wfs*. are theorems of L, where $\mathcal{A}$ and $\mathcal{B}$ are any *wfs*. of L.
   - (a) $(\mathcal{A} \to (\sim(\sim\!\mathcal{A})))$;
   - (b) $((\mathcal{B} \to \mathcal{A}) \to ((\sim\!\mathcal{A}) \to (\sim\!\mathcal{B})))$;
   - (c) $(((\mathcal{A} \to \mathcal{B}) \to \mathcal{A}) \to \mathcal{A})$;
   - (d) $(\sim(\mathcal{A} \to \mathcal{B}) \to (\mathcal{B} \to \mathcal{A}))$.

4 Let L' be the formal deductive system which differs from L only in having the axiom scheme $(L'3)$ $((\sim\!\mathcal{A} \to \sim\!\mathcal{B}) \to ((\sim\!\mathcal{A} \to \mathcal{B}) \to \mathcal{A}))$ in place of $(L3)$. Show that, for any *wfs*. $\mathcal{A}$ and $\mathcal{B}$ of L (and so of L'):
   - (i) $\underset{L}{\vdash} ((\sim\!\mathcal{A} \to \sim\!\mathcal{B}) \to ((\sim\!\mathcal{A} \to \mathcal{B}) \to \mathcal{A}))$,

and
   - (ii) $\underset{L'}{\vdash} ((\sim\!\mathcal{A} \to \sim\!\mathcal{B}) \to (\mathcal{B} \to \mathcal{A}))$.

Deduce that a *wf*. is a theorem of L if and only if it is a theorem of L'.

5   The rule *HS* is an example of a legitimate additional rule of deduction for *L*. Is the following rule legitimate in the same sense: from the *wfs.* $\mathcal{B}$ and $(\mathcal{A} \to (\mathcal{B} \to \mathcal{C}))$, deduce $(\mathcal{A} \to \mathcal{C})$?

2.2 The Adequacy Theorem for *L*

It is not a very rewarding exercise just to prove that certain particular *wfs.* of *L* are theorems of *L* as we have done in Proposition 2.11. As part (*b*) shows, it is sometimes difficult to know how to proceed, and the end product can be very complex and far from an intuitive proof. However, this aspect of *L* need not cause us concern. The reason for defining *L* in the first place was as an attempt to construct a formal system which reflected (by analogy) our intuitive ideas of deduction, validity and truth, and thereby to try to learn something about those intuitive ideas.

Chapter 1 gave us a notion of 'logical truth', namely that of tautology. It would be reasonable to hope that these logical truths will correspond to the theorems of *L*, and to attempt to construct *L* with this end in view. The remainder of this chapter is devoted to showing that *L* has this property. This procedure will also give us some insight into the nature and properties of formal systems in general which will be useful in later chapters.

Though the symbols of the language of *L* are being thought of purely as formal symbols, *L* was defined in such a way that we could interpret the *wfs.* of *L* as statement forms and that then each truth function is represented by some *wf.* Thus, although we cannot talk of assigning truth values to the symbols of *L* in precisely the same way as in Chapter 1, we can define an analogous procedure.

Definition 2.12

A *valuation* of *L* is a function v whose domain is the set of *wfs.* of *L* and whose range is the set $\{T, F\}$ such that, for any *wfs.* $\mathcal{A}$, $\mathcal{B}$ of *L*,

(i) $v(\mathcal{A}) \neq v(\sim\mathcal{A})$,

and

(ii) $v(\mathcal{A} \to \mathcal{B}) = F$ if and only if $v(\mathcal{A}) = T$ and $v(\mathcal{B}) = F$.

Note that an arbitrary 'assignment of truth values' to the symbols $p_1, p_2, \ldots$ of *L* will yield a valuation, as each *wf.* of *L* will (as a statement form) take one of the two truth values under such an assignment. (i) and (ii) will then obviously be satisfied.

Definition 2.13

A *wf.* $\mathcal{A}$ of *L* is a *tautology* if for every valuation v, $v(\mathcal{A}) = T$. This is the same as regarding $\mathcal{A}$ as a statement form and applying the previous definition.

▷ We shall prove that a *wf.* of *L* is a theorem of *L* if and only if it is a tautology. Already we are able to prove the implication one way.

Proposition 2.14 (The Soundness Theorem)

Every theorem of *L* is a tautology.

Proof. Let $\mathscr{A}$ be a theorem of *L*. The proof is by induction on the number of *wfs.* of *L* in a sequence of *wfs.* which constitutes a proof of $\mathscr{A}$ in *L*.

For the base step, suppose that the proof of $\mathscr{A}$ has only one *wf.* in it, namely $\mathscr{A}$. Then $\mathscr{A}$ must be an axiom of *L*. All the axioms of *L* are tautologies. The verification of this is by constructing truth tables, and is left as an exercise for the reader.

Now suppose that the proof of $\mathscr{A}$ contains *n* *wfs.*, where $n > 1$, and suppose as induction hypothesis that all theorems of *L* which have proofs in fewer than *n* steps are tautologies. Either $\mathscr{A}$ is an axiom, in which case $\mathscr{A}$ is a tautology, or $\mathscr{A}$ follows by *MP* from two previous *wfs.* in the proof. These two *wfs.* must have the forms $\mathscr{B}$ and $(\mathscr{B} \rightarrow \mathscr{A})$. But $\mathscr{B}$ and $(\mathscr{B} \rightarrow \mathscr{A})$ are theorems of *L* with proof sequences containing fewer than *n* *wfs.* (namely the proof of $\mathscr{A}$ truncated appropriately). Hence $\mathscr{B}$ and $(\mathscr{B} \rightarrow \mathscr{A})$ are tautologies, by the induction hypothesis, and so, by Proposition 1.9, $\mathscr{A}$ is a tautology.

Hence, by the principle of mathematical induction, every theorem of *L* is a tautology.

▷ To prove the converse of this, we need two new ideas: *extensions* of *L* and *consistency*.

L has three axiom schemes, and these are the starting points for the proofs of theorems. What would happen if we added another axiom scheme, or just another single axiom? We would have more to start from, so in general we would expect to be able to prove more theorems. All *wfs.* which were previously theorems would remain theorems, but perhaps some *wfs.* which were not theorems would become theorems. In fact, new theorems will arise if and only if the new set of axioms includes a *wf.* which was not previously a theorem. (The reader may readily supply a proof of this.)

Definition 2.15

An *extension* of *L* is a formal system obtained by altering or enlarging the set of axioms so that all theorems of *L* remain theorems (and new theorems are possibly introduced).

One need only look in some of the many existing logic textbooks to see that it is possible to replace our axiom schemes (*L*1), (*L*2), and (*L*3) by others in such a way that the class of theorems is unchanged. For example (see Exercise 2.4), (*L*3) may be replaced by the scheme:

$$((\sim\mathscr{A} \to \sim\mathscr{B}) \to ((\sim\mathscr{A} \to \mathscr{B}) \to \mathscr{A}))$$

without altering the class of theorems.

Note. It is possible for a formal system to be an extension of *L*, but to have no axioms in common with *L*.

▷ If we were to extend *L* to systems with more and more theorems, the more likely it will be that there will be some *wf.* $\mathscr{A}$ such that both $\mathscr{A}$ and $(\sim\mathscr{A})$ are theorems. Such a situation will clearly be undesirable.

Definition 2.16

An extension of *L* is *consistent* if for no *wf.* $\mathscr{A}$ of *L* are both $\mathscr{A}$ and $(\sim\mathscr{A})$ theorems of the extension.

This definition would of course be irrelevant if *L* itself were not consistent.

Proposition 2.17

L is consistent.

Proof. Suppose that *L* is not consistent, i.e. that there is a *wf.* $\mathscr{A}$ such that $\vdash_{L} \mathscr{A}$ and $\vdash_{L} (\sim\mathscr{A})$. Then by Proposition 2.14 (The Soundness Theorem), both $\mathscr{A}$ and $(\sim\mathscr{A})$ are tautologies. This is impossible since if $\mathscr{A}$ is a tautology then $(\sim\mathscr{A})$ is a contradiction and if $(\sim\mathscr{A})$ is a tautology then $\mathscr{A}$ is a contradiction. Thus *L* must be consistent.

Proposition 2.18

An extension *L** of *L* is consistent if and only if there is a *wf.* which is not a theorem of *L**.

Proof. Let *L** be consistent. Then, for any *wf.* $\mathscr{A}$, either $\mathscr{A}$ or $(\sim\mathscr{A})$ is not a theorem (both cannot be theorems).

Conversely, suppose that *L** is not consistent. We show that there is no *wf.* which is not a theorem of *L**, i.e. that every *wf.* is a theorem of *L**. Let $\mathscr{A}$ be any *wf.* *L** is not consistent, so $\vdash_{L^*} \mathscr{B}$ and $\vdash_{L^*} (\sim\mathscr{B})$ for some *wf.* $\mathscr{B}$. Now $\vdash_{L} (\sim\mathscr{B} \to (\mathscr{B} \to \mathscr{A}))$, by Proposition 2.11. So $\vdash_{L^*} (\sim\mathscr{B} \to (\mathscr{B} \to \mathscr{A}))$, since *L** is an extension of *L*. Now applying *MP* twice, we get $\vdash_{L^*} \mathscr{A}$. Thus every *wf.* is a theorem of *L**, as required.

Two aspects of this proposition should be emphasised. They are:

(a) In an inconsistent extension of L, every *wf.* is a theorem. Whenever we make use of extensions of L we must be careful about consistency, for the value of a system in which every *wf.* is a theorem is as slight as that of a system in which no *wfs.* are theorems.

(b) The sufficient condition for consistency given by the proposition is a surprisingly weak one, namely that there be a single *wf.* which is not a theorem. This is surprising because in any consistent system there will certainly be many *wfs.* which are not theorems, since the negations of all theorems, for example, are non-theorems.

Now we come to a proposition which appears insignificant and technical, but we shall use it time and again in proofs of later results. It describes one circumstance in which a consistent extension can be obtained.

Proposition 2.19

Let L^* be a consistent extension of L and let $\mathscr{A}$ be a *wf.* of L which is not a theorem of L^*. Then L^{**} is also consistent, where L^{**} is the extension of L obtained from L^* by including $(\sim\mathscr{A})$ as an additional axiom.

Proof. Let $\mathscr{A}$ be a *wf.* of L which is not a theorem of L^*. Suppose that L^{**} is inconsistent. Then for some *wf.* $\mathscr{B}$, $\vdash_{L^{**}} \mathscr{B}$ and $\vdash_{L^{**}} (\sim\mathscr{B})$. Now it follows just as in the proof of Proposition 2.18 that $\vdash_{L^{**}} \mathscr{A}$. But L^{**} differs from L^* only in having $(\sim\mathscr{A})$ as an additional axiom, so $\vdash_{L^{**}} \mathscr{A}$ is equivalent to $(\sim\mathscr{A}) \vdash_{L^*} \mathscr{A}$. (A proof in L^{**} is just a deduction from $(\sim\mathscr{A})$ in L^*.) So $\vdash_{L^*} ((\sim\mathscr{A}) \to \mathscr{A})$, by the Deduction Theorem. But $\vdash_{L} (((\sim\mathscr{A}) \to \mathscr{A}) \to \mathscr{A})$, by Proposition 2.11, so $\vdash_{L^*} (((\sim\mathscr{A}) \to \mathscr{A}) \to \mathscr{A})$. Now by *MP*, we get

$$\vdash_{L^*} \mathscr{A}.$$

But this contradicts the hypothesis that $\mathscr{A}$ is not a theorem of L^*. Hence L^* must be consistent.

▷ There is obviously a limit somewhere to the *wfs.* which we can include as additional axioms in an extension of L while maintaining consistency. Attaining this limit is the object of the next proposition, but first let us describe the situation in a definition.

Definition 2.20

An extension of L is *complete* if for each *wf.* $\mathscr{A}$, either $\mathscr{A}$ or $(\sim\mathscr{A})$ is a theorem of the extension.

Remarks. (*a*) *L* is far from being complete. For example, p_1 is a *wf.* of *L*, and neither p_1 nor $(\sim p_1)$ are theorems of *L*.

(*b*) Any inconsistent extension of *L* is obviously complete, by Proposition 2.18.

(*c*) If L^c is a consistent complete extension of *L* then any further extension of *L* in which the class of theorems extends the class of theorems of L^c is inconsistent. For let $\mathscr{A}$ not be a theorem of L^c. Then $(\sim\mathscr{A})$ is a theorem of L^c. Hence if $\mathscr{A}$ is a theorem of a further extension, so is $(\sim\mathscr{A})$, and so this further extension cannot be consistent.

Proposition 2.21

Let L^* be a consistent extension of *L*. Then there is a consistent complete extension of L^*.

(Note that we have generalised our terminology here. An extension of L^* is obtained by altering or enlarging the set of axioms of L^* so as to enlarge the set of theorems. See Definition 2.15.)

Proof. Let $\mathscr{A}_0, \mathscr{A}_1, \mathscr{A}_2, \ldots$ be an enumeration of all the *wfs.* of *L*. This can be constructed in several ways – the reader is recommended to try to produce a method for generating such a list. (Familiarity with constructions involving infinite countable sets is useful in doing this.)† We shall build a sequence $J_0, J_1, J_2, \ldots$ of extensions of L^* as follows. Let

$$J_0 = L^*.$$

If

$$\vdash_{J_0} \mathscr{A}_0, \quad \text{let } J_1 = J_0.$$

If not $\vdash_{J_0} \mathscr{A}_0$, then add $(\sim\mathscr{A}_0)$ as a new axiom to obtain J_1 from J_0.

In general, for $n \geq 1$, to construct J_n from J_{n-1}: if $\vdash_{J_{n-1}} \mathscr{A}_{n-1}$, then $J_n = J_{n-1}$, and if not $\vdash_{J_{n-1}} \mathscr{A}_{n-1}$, then let J_n be the extension of J_{n-1} obtained by adding $(\sim\mathscr{A}_{n-1})$ as a new axiom.

L^* is consistent, i.e. J_0 is consistent, by assumption. For $n \geq 1$, if J_{n-1} is consistent, then J_n is consistent, by Proposition 2.19. Hence by induction, each J_n is consistent ($n \geq 0$). Now define *J* to be that extension of L^* which has as its axioms all the *wfs.* which are axioms of at least one of the J_n.

Now we show that *J* is consistent. Suppose the contrary. Then there is a *wf.* $\mathscr{A}$ such that $\vdash_J \mathscr{A}$ and $\vdash_J (\sim\mathscr{A})$. Now the proofs in *J* of $\mathscr{A}$ and $(\sim\mathscr{A})$

† See the Appendix.

are finite sequences of *wfs.*, so each proof can contain instances of only finitely many of the axioms of J. Therefore there must exist n which is large enough so that all these axioms which are used are axioms of J_n. It follows that $\vdash_{J_n} \mathscr{A}$ and $\vdash_{J_n} (\sim \mathscr{A})$. This contradicts the consistency of J_n, and so J must be consistent.

It remains to show that J is complete. Let $\mathscr{A}$ be a *wf.* of L. $\mathscr{A}$ must appear in the list $\mathscr{A}_0, \mathscr{A}_1, \mathscr{A}_2, \ldots$, say $\mathscr{A}$ is $\mathscr{A}_k$. If $\vdash_{J_k} \mathscr{A}_k$, then certainly $\vdash_{J} \mathscr{A}_k$, since J is an extension of J_k. If not $\vdash_{J_k} \mathscr{A}_k$, then according to the construction of J_{k+1}, $(\sim \mathscr{A}_k)$ is an axiom of J_{k+1}, and so $\vdash_{J_{k+1}} (\sim \mathscr{A}_k)$. This implies that $\vdash_{J} (\sim \mathscr{A}_k)$. So in any case we have $\vdash_{J} \mathscr{A}$ or $\vdash_{J} (\sim \mathscr{A})$, and so J is complete.

Proposition 2.22

If L^* is a consistent extension of L then there is a valuation in which each theorem of L^* takes value T.

Proof. Define v on *wfs.* of L by:

$$v(\mathscr{A}) = T \text{ if } \vdash_{J} \mathscr{A},$$

and

$$v(\mathscr{A}) = F \text{ if } \vdash_{J} (\sim \mathscr{A}),$$

where J is a consistent complete extension of L^* as given in the proof of Proposition 2.21. Note that v is defined on all *wfs.* since J is complete. Now $v(\mathscr{A}) \neq v(\sim \mathscr{A})$, for any *wf.* $\mathscr{A}$, since J is consistent, and it remains to show that $v(\mathscr{A} \to \mathscr{B}) = F$ if and only if $v(\mathscr{A}) = T$ and $v(\mathscr{B}) = F$. Suppose first that $v(\mathscr{A}) = T$, $v(\mathscr{B}) = F$ and that $v(\mathscr{A} \to \mathscr{B}) = T$. Then $\vdash_{J} \mathscr{A}$, $\vdash_{J} (\sim \mathscr{B})$ and $\vdash_{J} (\mathscr{A} \to \mathscr{B})$. It follows that $\vdash_{J} \mathscr{B}$, by *MP*, and the consistency of J is contradicted. Hence $v(\mathscr{A}) = T$ and $v(\mathscr{B}) = F$ imply $v(\mathscr{A} \to \mathscr{B}) = F$. Conversely, suppose that $v(\mathscr{A} \to \mathscr{B}) = F$ and either $v(\mathscr{A}) = F$ or $v(\mathscr{B}) = T$. Then $\vdash_{J} (\sim(\mathscr{A} \to \mathscr{B}))$ and either $\vdash_{J} (\sim \mathscr{A})$ or $\vdash_{J} \mathscr{B}$. Now

$$\vdash_{J} (\sim \mathscr{A} \to (\sim \mathscr{B} \to \sim \mathscr{A})),$$

and

$$\vdash_{J} (\mathscr{B} \to (\mathscr{A} \to \mathscr{B})),$$

so by *MP*,

$$\vdash_{J} (\sim \mathscr{B} \to \sim \mathscr{A}) \text{ or } \vdash_{J} (\mathscr{A} \to \mathscr{B}).$$

Now $\vdash_J ((\sim \mathcal{B} \to \sim \mathcal{A}) \to (\mathcal{A} \to \mathcal{B}))$, so in either case we have $\vdash_J (\mathcal{A} \to \mathcal{B})$, contradicting the consistency of J. Hence $v(\mathcal{A} \to \mathcal{B}) = F$ implies $v(\mathcal{A}) = T$ and $v(\mathcal{B}) = F$. Therefore v is a valuation.

Now let $\mathcal{A}$ be a theorem of L^*. Then $\vdash_J \mathcal{A}$ since J is an extension of L^*. Therefore $v(\mathcal{A}) = T$.

$\triangleright$ Now we are in a position to prove our desired result.

Proposition 2.23 (The Adequacy Theorem for L)

If $\mathcal{A}$ is a *wf.* of L and $\mathcal{A}$ is a tautology, then $\vdash_L \mathcal{A}$.

Proof. Let $\mathcal{A}$ be a *wf.* of L and a tautology, and suppose that $\mathcal{A}$ is not a theorem of L. Then the extension L^*, obtained by including $(\sim \mathcal{A})$ as a new axiom, is consistent, by Proposition 2.19. Therefore there is a valuation v which gives every theorem of L^* the value T. In particular, $v(\sim \mathcal{A}) = T$. But $v(\mathcal{A}) = T$, because $\mathcal{A}$ is a tautology, and so we have a contradiction. Hence $\mathcal{A}$ is a theorem of L.

$\triangleright$ Now we have verified that the formal system L has the principal property which it ought to have, namely that the *wf*s. which can be proved within it are precisely those which are 'logically true'. The axioms and rule of deduction in L characterise logical deduction, at least in this context. The value of our study of L lies in this and not in detailed study of *wf*s., proofs and theorems of L.

The ideas and methods which we have used in order to prove the Adequacy Theorem are quite strong and will have other applications, though their use here was merely as tools for a particular purpose. There are many other proofs in the literature of the Adequacy Theorem, some using totally different methods. We have used this method because we can apply more or less the same procedure later to prove a corresponding Adequacy Theorem for the more complicated formal system of predicate calculus. In fact our whole discussion of the system L has been in order to introduce the ideas of the subject rather than to study the statement calculus for its own sake. As far as that is concerned, the methods of Chapter 1 provide us with sufficient insight into the subject. Truth tables give us all the information we need about particular statement forms and argument forms, and by using them we can effectively differentiate between tautologies, contradictions and other statement forms. The consequence of this for L is important and useful, so we include it as a proposition.

Proposition 2.24

L is *decidable*, i.e. there is an effective method for deciding, given any *wf.* of L, whether it is a theorem of L.

Proof. To tell whether a *wf.* $\mathscr{A}$ is a theorem of L, just consider it as a statement form and construct its truth table. It is a theorem if and only if it is a tautology.

Remark 2.25

This renders unnecessary any further construction of proofs in L. Truth tables give a mechanical, if not always quick, method of showing whether a given *wf.* is a theorem of L. But of course we did not know this until now, so the proof of Proposition 2.11, for example, cannot now be changed, as its result was required in the proof of the Adequacy Theorem.

Exercises

6 Prove that every axiom of L is a tautology.

7 Let $\mathscr{A}$ be a *wf.* of L and let L^+ be the extension of L obtained by including $\mathscr{A}$ as a new axiom. Prove that the set of theorems of L^+ is different from the set of theorems of L if and only if $\mathscr{A}$ is not a theorem of L.

8 Let $\mathscr{A}$ be the *wf.* $((\sim p_1 \to p_2) \to (p_1 \to \sim p_2))$. Show that L^+, obtained by including this $\mathscr{A}$ as a new axiom, has a larger set of theorems than L. Is L^+ a consistent extension of L?

9 Prove that if $\mathscr{B}$ is a contradiction then $\mathscr{B}$ cannot be a theorem of any consistent extension of L.

10 Let L^{++} be the extension of L obtained by including as a fourth axiom *scheme*:

$$((\sim \mathscr{A} \to \mathscr{B}) \to (\mathscr{A} \to \sim \mathscr{B})).$$

Show that L^{++} is inconsistent. (Hint: see Chapter 1 exercise 7.)

11 Let J be a consistent complete extension of L, and let $\mathscr{A}$ be a *wf.* of L. Show that the extension of J obtained by including $\mathscr{A}$ as an additional axiom is consistent if and only if $\mathscr{A}$ is a theorem of J.

12 Let $\mathscr{A}$ be a *wf.* of L in which the statement letters $p_1, \ldots, p_n$ occur, and let $\mathscr{A}_1, \mathscr{A}_2, \ldots, \mathscr{A}_n$ be any *wfs.* of L. Let $\mathscr{B}$ be the *wf.* of L obtained by substituting $\mathscr{A}_i$ for each occurrence of p_i in $\mathscr{A}$ $(1 \le i \le n)$. Prove that if $\mathscr{A}$ is a theorem of L then $\mathscr{B}$ is a theorem of L.

3
Informal predicate calculus

3.1 Predicates and quantifiers

In Chapter 1 we analysed sentences and arguments, breaking them down into constituent simple statements, regarding these simple statements as the building blocks. By this means we were able to discover something of what makes a valid argument. However there are arguments which are not susceptible to such a treatment. For example, let us write down one of the examples of Chapter 1 in a slightly different form:

> All men are mortal
> Socrates is a man
> ∴ Socrates is mortal.

This we would regard intuitively as an example of a valid argument, but if we try to symbolise the form of the argument as we did in Chapter 1, what we get is *p*, *q*, ∴ *r*. According to Chapter 1, this is not a valid argument form.

Validity in this case depends not upon the relationships of the premisses and conclusion as simple statements, but upon relationships between *parts* of the statements involved and upon the forms of the statements themselves. If we wished to make this clearer by finding a corresponding 'argument form', it would have to look like this:

> All *A*s are *B*
> *C* is an *A*
> ∴ *C* is *B*.

There are two points to be dealt with. First, the general nature of the premiss 'All *A*s are *B*', and second, the use of symbols to represent parts of simple statements. These points correspond respectively to the ideas of 'quantifier' and 'predicate'. Every simple statement in English has a subject and a predicate, each of which may consist of a single word, a short phrase or a whole clause. Putting it very roughly, the subject is the thing about which the statement is making an assertion, and the predicate refers to a 'property' which the subject has.

Example 3.1

In each of the following statements the subject is underlined and the remainder is the predicate.

(*a*)   <u>Socrates</u> is a man.

(*b*)   <u>I</u> write books.

(*c*)   <u>The number whose square is −1</u> is not real.

(*d*)   <u>The world</u> owes everyone a living.

It is convenient to represent predicates by capital letters $A, B, C, \ldots$, and subjects by small letters, and thereby to symbolise statements such as the above as follows:

(*a*) $A(s)$ may stand for 'Socrates is a man', where A is a predicate letter standing for 'is a man' and s stands for Socrates. We could then symbolise 'Napoleon is a man' by $A(n)$, with n standing for Napoleon.

(*b*) $B(i)$ may stand for 'I write books' in a similar way.

(*c*) In this case our predicate is a negation, so we have a choice. R could be chosen to mean 'is not real', so that the statement would have the form $R(j)$, where j stands for the number whose square is −1. Or we could choose S to mean 'is real', so that the statement would be $(\sim S(j))$.

(*d*) $L(w)$, similarly to (*a*) and (*b*).

It should be clear that compound statements can also be translated into symbols in this way, just by symbolising all the constituent simple statements.

▷ Now what about statements such as 'all men are mortal'? We need something better than a subject–predicate analysis, because the meaning of the statement depends on the force of the word 'all'. Consider another example:

> Every integer has a prime factor.

In ordinary mathematical symbolism we might write this as:

> For all x, if x is an integer then x has a prime factor.

Using the kind of symbolic language just introduced we could write this as:

> For all x, $(I(x) \to P(x))$,

where $I(x)$ stands for 'x is an integer' and $P(x)$ stands for 'x has a prime factor'.

Similarly, if we introduce predicate symbols A and M for 'is a man' and 'is mortal' respectively, then 'all men are mortal' may be written:

> For all x, $(A(x) \to M(x))$.

The phrase 'for all x' is called a *universal quantifier* and is translated into symbols as $(\forall x)$. Note that when we write $(\forall x)(A(x) \rightarrow M(x))$ there is no assumption about the nature of the object x. The implication is asserted 'for every object x in the universe'. If x is a man, then x is mortal. For any x which is not a man, whether that particular x is mortal is irrelevant. The implication is true because its first part is false (see the truth table for $\rightarrow$).

The introduction of the symbol x ought not to cause confusion, even though it does not appear in the original English sentence. Its use is merely a mathematical abbreviation, and clearly 'all men are mortal' may be translated into symbols also as:

For all y, $(A(y) \rightarrow M(y))$.

We shall use the letters x and y as variables, as indeterminate subjects. When they are used as above in statements starting with quantifiers, they are said to be *bound* variables.

There is another kind of quantifier which is at first sight needed to translate ordinary English statements into symbols. Consider the sentence 'some pigs have wings'. A rephrasing of this would be 'there is at least one pig which has wings', or, using a device similar to that above:

There exists at least one object x such that
x is a pig and x has wings.

The phrase 'there exists at least one object x such that' is called an *existential quantifier*, and is translated into symbols as $(\exists x)$. The sentence may now be written

$(\exists x)(P(x) \wedge W(x))$,

where $P(x)$ and $W(x)$ mean 'x is a pig' and 'x has wings' respectively.

More generally, if A is a symbol which stands for a predicate, then we can write meaningfully $(\forall x)A(x)$ and $(\exists x)A(x)$. The first means 'every object has the property determined by A', and the second means 'there is some object which has the property determined by A'.

Example 3.2

Translate into symbols.
- (a) Not all birds can fly.
- (b) Anyone can do that.
- (c) Some people are stupid.
- (d) There is an integer which is greater than every other integer.

Answers: (of course there may be different ways).
- (a) $\sim(\forall x)(B(x) \rightarrow F(x))$.
- (b) $(\forall x)(M(x) \rightarrow C(x))$.

It helps to go through the steps as we did above. 'Anyone can do that' means 'all people can do that', and just as before this means 'for all x, if x is a person then x can do that'. $M(x)$ stands for 'x is a person' and $C(x)$ stands for 'x can do that'.

(c) $(\exists x)(M(x) \wedge S(x))$.

(d) $(\exists x)(I(x) \wedge (\forall y)(I(y) \rightarrow x \geq y))$.

We shall find that these illustrate a common, but not universal, pattern. The universal quantifier is very often followed by an implication, because a universal statement is most often of the form 'given any x, if it has property A then it also has property B'. The existential quantifier is very often followed by a conjunction, because an existential statement is most often of the form 'there exists an x with property A, which also has property B'.

Let us consider (a) further. It is well known that this assertion is true, and we can justify it by the instances of ostriches, kiwis or penguins. Intuitively we are justifying 'not all birds can fly' by justifying 'there is a bird which cannot fly'. Here is an important connection between the two quantifiers, for a moment's thought will convince one that the statements above have the same meaning. Let us translate them into symbols.

(i) $\sim(\forall x)(B(x) \rightarrow F(x))$,

and

(ii) $(\exists x)(B(x) \wedge \sim F(x))$.

To compare these more closely, let us transform the first into

$$\sim(\forall x)(\sim B(x) \vee F(x)),$$

according to the rules of Chapter 1, and further into

$$\sim(\forall x) \sim (B(x) \wedge \sim F(x)).$$

Now this has a similar form to (ii), but with $\sim(\forall x)\sim$ in place of $(\exists x)$.

Consideration of examples like this enables us to see intuitively that the two sentences:

(i)   It is not the case that all x's do not have property P,

(ii)   There exists an x which has property P,

have the same meaning, whatever property P stands for.

Example 3.3

Translate into symbols each of the following, first using no universal quantifiers, then using no existential quantifiers.

(a)   All birds can fly.

(b)   No man is an island.

(c)   Some numbers are not rational.

Answers:
   (a) $\sim(\exists x)(B(x) \wedge \sim F(x))$
          $(\forall x)(B(x) \rightarrow F(x))$.
   (b) $\sim(\exists x)(M(x) \wedge I(x))$
          $(\forall x)(M(x) \rightarrow \sim I(x))$.
   (c) $(\exists x)(N(x) \wedge \sim R(x))$
          $\sim(\forall x)(N(x) \rightarrow R(x))$.
(There are, of course, other equivalent ways.)

▷ Now that we know how to translate into symbols, how does this help in deciding the relationships between statements or the validity of arguments? It is not possible to extend the use of truth tables, because our sentences now are not truth functional. The use of variables and quantifiers means that the truth value of a sentence does not depend as it did before merely on the truth values of the component parts, nor do the parts always have truth values. In particular, it makes no sense to talk of the truth value of a part of a sentence containing a variable which has no quantifier, for example 'x is a bird', or $B(x)$ in our previous example.

Exercises

1 Translate into symbols the following statements, using quantifiers, variables and predicate symbols.
   (a)   Not every function has a derivative.
   (b)   There is a function which is continuous but has no derivative.
   (c)   If some trains are late then all trains are late.
   (d)   Every number is either odd or even.
   (e)   No number is both odd and even.
   (f)   Some people hate everyone.
   (g)   Elephants are heavier than mice.
2 Translate each of the following statements into symbols, first using no existential quantifiers, and second using no universal quantifiers.
   (a)   Not all cars have three wheels.
   (b)   Some people are either lazy or stupid.
   (c)   No mouse is heavier than any elephant.
   (d)   Every number either is negative or has a square root.
3 Pick out any pairs of statements from Exercises 1 and 2 which have the same meanings.

3.2 First order languages

One method for analysing statements and arguments involving quantifiers is syllogistic logic. This subject has a long pedigree, starting with Aristotle and stretching to the present day. The interested reader may pursue it further in other books (e.g. Copi), as we shall not go into it here. Its basis is the study of a small number of particular argument forms which are known intuitively to be valid, for example:

> All As are B
> No C is B
> $\therefore$ Some C is not A.

The purpose is, given a particular argument, to express it by means of one or more of these basic argument forms, thereby demonstrating its validity. Modern logicians and mathematicians have found this system too restrictive, and so have searched for a different method of analysis. What has been developed, and what we shall study, has the great mathematical merit of leading on to new and rich areas of study which are not even contemplated in syllogistic logic.

We construct a formal system. It is a more complicated system than that in Chapter 2, as is only to be expected, but the principles on which it is built are the same. We must first describe a formal language, giving the alphabet of symbols and the rule for the construction of well-formed formulas. In this we are guided by our experience above in translating ordinary sentences into symbols, since our purpose is to make the formal language such that we can translate such ordinary sentences into well-formed formulas of the system, and to reflect the logical properties of sentences in the properties of *wfs.* within the system. It should be made clear again at this point that the symbols of the formal language are to have no meanings or properties other than those specified within the formal system. They may on occasions be interpreted in different ways, but these interpretations will not be part of the system.

The alphabet of symbols is the following:

$x_1, x_2, \ldots$	variables,
$a_1, a_2, \ldots$	individual constants,
$A_1^1, A_2^1, \ldots; A_1^2, A_2^2, \ldots; A_1^3, A_2^3, \ldots; \ldots$	predicate letters,
$f_1^1, f_2^1, \ldots; f_1^2, f_2^2, \ldots; f_1^3, f_2^3, \ldots; \ldots$	function letters,
$(,),\,,$	punctuation,
$\sim, \rightarrow$	connectives,
$\forall$	quantifier.

Remarks 3.4

(a) The individual constants are included so that we shall have formulas in our language which can be interpreted as statements about particular things. For example, the statement 'Socrates is a man' could be an interpretation of the formula $A_1^1(a_1)$.

(b) There is a list of lists of predicate letters. The first is a list of one place predicate letters intended to be interpreted by one place predicates (like 'is a man'). The next is a list of two place predicate letters, to stand for relations or two place predicates (like 'is the father of'). And so on. (An example of a three place predicate in ordinary language is 'are collinear' in the statement 'the points A, B and C are collinear'.)

(c) We have not seen any letters representing functions in any of our informal translations into symbols up till now. The idea of a function is so fundamental in mathematics that it is advantageous to allow letters for functions in the formal language. This is because the intended interpretations of the symbols are in the main mathematical. Of course, a function is a special kind of relation, and indeed it would be sufficient to have symbols just for relations (predicate letters), but we are not interested merely in making the alphabet of symbols as small as possible. Another criterion that we are applying is intuitive clarity. Later when we discuss specific mathematical systems, we shall see quite clearly the effect of including function letters. Just as for predicate letters, there are separate lists of function letters, intended to stand for functions with different numbers of arguments. In both cases, the superscript number indicates the number of places or arguments.

(d) There is only one quantifier in the language, the universal quantifier. We have seen that the existential quantifier can be defined in terms of the universal quantifier, so we need have only one of them, in a similar way to our inclusion of only $\sim$ and $\rightarrow$ as our connectives.

(e) There are infinite lists of symbols, in spite of the fact that we shall allow only finite combinations of them as well-formed formulas (see below). We need a *potentially* infinite number of symbols in order to keep our language as general as possible. In applications we shall specify interpretations for only some of these symbols, and some applications may need more symbols than others, so we do not wish to put an upper bound on the number of interpretable symbols.

▷ In general, a *first order language* $\mathscr{L}$ will have as its alphabet of symbols:
  variables $x_1, x_2, \ldots$,
  some (possibly none) of the individual constants $a_1, a_2, \ldots$,
  some (possibly none) of the predicate letters A_i^n,
  some (possibly none) of the function letters f_i^n,
  the punctuation symbols (,) and , ,
  the connectives $\sim$ and $\rightarrow$,
  the quantifier $\forall$.
It is clear that there are many different first order languages, depending on the symbols which are included. In most of our work we shall not specify which language we employ, and the results which we obtain are

therefore applicable to any language. The significance of the term 'first order' will emerge later on, but it is connected with the use of the universal quantifier.

To specify a formal language completely, we have to say what a well-formed formula is, but before we do that, let us consider some examples.

Example 3.5

(a) If we wish our first order language to be appropriate for statements about the arithmetic of natural numbers, we might take $\mathscr{L}$ to have (along with variables, punctuation, connectives and quantifier) the symbols:

a_1, to stand for 0,
A_1^2, to stand for =,
f_1^1, to stand for the successor function,
f_1^2, to stand for +,
f_2^2, to stand for ×.

Then, for example,

$$A_1^2(f_1^2(x_1, x_2), f_2^2(x_1, x_2))$$

would be interpreted as '$x_1 + x_2 = x_1 x_2$'.

(b) If we wish our first order language to be appropriate for statements about groups, we could take $\mathscr{L}$ to have (along with variables, punctuation, connectives and quantifier) the symbols:

a_1, to stand for the identity element,
A_1^2, to stand for =,
f_1^1, to stand for the function which takes each element to its inverse,
f_1^2, to stand for the group binary operation.

Then, for example,

$$A_1^2(f_1^2(x_1, f_1^1(x_1)), a_1)$$

would be interpreted as '$x_1 x_1^{-1} = $ identity'.

Definition 3.6

Before defining well-formed formulas, we need some preliminaries. Let $\mathscr{L}$ be a first order language. A *term* in $\mathscr{L}$ is defined as follows.

(i) Variables and individual constants are terms.

(ii) If f_i^n is a function letter in $\mathscr{L}$, and $t_1, \ldots, t_n$ are terms in $\mathscr{L}$, then $f_i^n(t_1, \ldots, t_n)$ is a term in $\mathscr{L}$.

(iii) The set of all terms is generated as in (i) and (ii).

Terms are to be those expressions in the formal language which will be interpreted as objects, i.e. the things to which functions are applied, the things which have properties, the things about which assertions are made.

An *atomic formula* in $\mathscr{L}$ is defined by: if A_j^k is a predicate letter in $\mathscr{L}$ and $t_1, \ldots, t_k$ are terms in L, then $A_j^k(t_1, \ldots, t_k)$ is an atomic formula of $\mathscr{L}$.

Atomic formulas are the simplest expressions in the language which are to be interpreted as assertions, for example that a certain property holds of certain objects. The word 'atomic' of course means 'unable to be broken down'.

Atomic formulas, continuing the analogy which the terminology suggests, are what well-formed formulas are built up from. They combine according to logical rules, and hold a place corresponding to that of statement letters in our previous formal system.

A *well-formed formula* of $\mathscr{L}$ is defined by:

(i) Every atomic formula of $\mathscr{L}$ is a *wf*. of $\mathscr{L}$.

(ii) If $\mathscr{A}$ and $\mathscr{B}$ are *wfs*. of $\mathscr{L}$, so are $(\sim\mathscr{A})$, $(\mathscr{A} \to \mathscr{B})$ and $(\forall x_i)\mathscr{A}$, where x_i is any variable.

(iii) The set of all *wfs*. of $\mathscr{L}$ is generated as in (i) and (ii).

Remark 3.7

(*a*) The *wfs*. are built up from atomic formulas just as in the case of the system L, with the exception, of course, of the inclusion of the universal quantifier. If $\mathscr{A}$ is a *wf*. of $\mathscr{L}$, then so is $(\forall x_i)\mathscr{A}$, where x_i is any variable. Hence, as an example, if $A_1^1(x_2)$ is a *wf*. of $\mathscr{L}$, so is $(\forall x_1)A_1^1(x_2)$. Thus the quantifier need not have any connection with the *wf*. to which it is applied, though obviously we are likely to be concerned more with cases where the quantified variable does occur in the subsequent formula.

(*b*) As in the system L, we include only the connectives $\sim$ and $\to$. This is in order to simplify our proofs of properties of the language and of formal systems based on the language. It is for the same reason that we exclude the symbol $\exists$. However we shall find it convenient later to use the symbol $\exists$ and the connectives $\land$ and $\lor$ as *defined* symbols, following the intuitive guide given earlier.

$(\exists x_i)\mathscr{A}$ is an abbreviation for $(\sim((\forall x_i)(\sim\mathscr{A})))$,

$(\mathscr{A} \land \mathscr{B})$ is an abbreviation for $(\sim(\mathscr{A} \to (\sim\mathscr{B})))$,

$(\mathscr{A} \lor \mathscr{B})$ is an abbreviation for $((\sim\mathscr{A}) \to \mathscr{B})$.

Formulas which include these symbols are not strictly *wfs*. of the formal language, but they may be translated back into *wfs*. if necessary.

(*c*) The use of parentheses in well-formed formulas is precisely given in the definition. As we did with L, however, we shall sometimes omit parentheses, as long as no ambiguity is introduced. As before, a $\sim$ will be presumed to apply to the shortest possible subsequent *wf*., e.g.

$(\sim\!\mathscr{A} \to \mathscr{B})$ is an abbreviation for $((\sim\!\mathscr{A}) \to \mathscr{B})$. In a similar way we treat quantifiers, e.g. $((\forall x_i)\mathscr{A} \to \mathscr{B})$ is in fact a *wf.* with no parentheses omitted, and the difference should be noted between this *wf.* and the *wf.* $(\forall x_i)$ $(\mathscr{A} \to \mathscr{B})$. Let us introduce some terminology in relation to this.

Definition 3.8

In the *wf.* $(\forall x_i)\mathscr{A}$, we say that $\mathscr{A}$ is the *scope* of the quantifier. More generally, when $((\forall x_i)\mathscr{A})$ occurs as a subformula of a *wf.* $\mathscr{B}$, we say that the scope of this quantifier in $\mathscr{B}$ is $\mathscr{A}$.

An occurrence of the variable x_i in a *wf.* is said to be *bound* if it occurs within the scope of a $(\forall x_i)$ in the *wf.*, or if it is the x_i in a $(\forall x_i)$. If an occurrence of a variable is not bound it is said to be *free*.

Example 3.9

(a) In the *wf.* $(\forall x_1)A_1^1(x_2)$, the scope of the quantifier $(\forall x_1)$ is $A_1^1(x_2)$, the variable x_2 occurs free, and the variable x_1 occurs bound.

(b) $(\forall x_1)(\forall x_2)(A_1^2(x_1, x_2) \to A_1^1(x_2))$ is a *wf.* in which all occurrences of x_1 and x_2 are bound. The scope of the $(\forall x_1)$ is $(\forall x_2)(A_1^2(x_1, x_2) \to A_1^1(x_2))$, and the scope of the $(\forall x_2)$ is $(A_1^2(x_1, x_2) \to A_1^1(x_2))$.

(c) $(\forall x_1)(A_1^2(x_1, x_2) \to (\forall x_2)A_1^1(x_2))$ is a *wf.* in which x_1 occurs bound twice and x_2 occurs free once and bound twice. The scope of the $(\forall x_1)$ is $(A_1^2(x_1, x_2) \to (\forall x_2)A_1^2(x_2))$, and the scope of the $(\forall x_2)$ is $A_1^1(x_2)$.

▷ In the above we have been using symbols which are not part of the formal language, namely $\mathscr{A}$ and $\mathscr{B}$. The reader will recall a similar situation in Chapter 2. These letters are part of the ordinary mathematical language which we are using to describe and discuss the formal language, and they are allowed to stand for well-formed formulas (usually arbitrary) in the formal language. We shall sometimes also write $\mathscr{A}(x_1)$ or $\mathscr{B}(x_1, \ldots, x_n)$, for example, when we are interested in particular variables (or terms). Such expressions will often, but not always, indicate that the variables mentioned occur free in the *wf.* If x_i does occur free in $\mathscr{A}(x_i)$, then for any term t, $\mathscr{A}(t)$ will denote the result of substituting t for every free occurrence of x_i.

Example 3.10

Let us examine the use of quantified variables, and return to our previous example $(\forall x_1)A_1^1(x_2)$. Intuitively this is intended to be interpreted along the lines of 'whatever object x_1 stands for, the property determined by A_1^1 holds of the object that x_2 stands for'. The quantifier is clearly redundant, and other quantifiers in place of the $(\forall x_1)$ would be redundant in the same way. $(\forall x_3)A_1^1(x_2)$ would have the same intuitive interpretation, and so would $(\forall x_7)A_1^1(x_2)$. However, it is just as clear that the intuitive interpretation of $(\forall x_2)A_1^1(x_2)$ is essentially different from that of any of the above. It is 'whatever object x_2 stands for, the property determined by A_1^1 holds of that object'. And, without reference to intuitive interpretations, there is a purely formal difference between $(\forall x_2)A_1^1(x_2)$ and the others, and that is that the two variables which appear are the same.

We shall sometimes have occasion to replace variables by other variables or terms, and we shall want to do it in such a way that no difference will result in the intuitive interpretation. To use the same example, $(\forall x_1)A_1^1(x_2)$: if we replace x_2 by x_3 we get a *wf.* which has the same form of interpretation, and if we replace x_2 by x_1 we get a *wf.* which has a different form of interpretation. We say that x_3 is free for x_2 in $(\forall x_1)A_1^1(x_2)$ but that x_1 is not. More generally, if we replace x_2 by the term $f_1^2(x_1, x_3)$, say, we obtain $(\forall x_1)A_1^1(f_1^2(x_1, x_3))$, and there is an interaction between the quantifier $(\forall x_1)$ and its scope which was not present previously. So we say that $f_1^2(x_1, x_3)$ is not free for x_2 in $(\forall x_1)A_1^1(x_2)$. However, any term which does not involve the variable x_1 is free for x_2 in $(\forall x_1)A_1^1(x_2)$. Let us extend these ideas into a general definition.

Definition 3.11

Let $\mathscr{A}$ be any *wf.* of $\mathscr{L}$. A term t is *free for x_i in $\mathscr{A}$* if x_i does not occur free in $\mathscr{A}$ within the scope of a $(\forall x_j)$, where x_j is any variable occurring in t.

Roughly, this means (as above) that t may be substituted for every free occurrence of x_i in $\mathscr{A}$ without introducing any interactions with quantifiers in $\mathscr{A}$.

Example 3.12

We have seen some examples above. More complicated is

$$((\forall x_1)A_1^2(x_1, x_2) \rightarrow (\forall x_3)A_2^2(x_3, x_1)).$$

Here, for example, $f_1^2(x_1, x_4)$ is not free for x_2, $f_2^2(x_2, x_3)$ is free for x_2, x_2 is free for x_1 (note that x_1 occurs only once free), and $f_4^2(x_1, x_3)$ is not free for x_1.

Remark 3.13

For any *wf.* $\mathscr{A}$, and any variable x_i (whether it occurs free in $\mathscr{A}$ or not), x_i is free for x_i in $\mathscr{A}$.

▷ We have now described the formal language which we shall be using and developing. We shall proceed, as we did in Chapter 2, to give axioms and rules of deduction to complete the specifications of formal systems, which we shall call systems of *First Order Predicate Calculus*. Before doing this, however, it is appropriate to investigate the vague word 'interpretation' which we have used several times already, and try to make it precise.

Exercises

4   Let $\mathcal{L}$ be a first order language whose alphabet of symbols contains no function letters. Describe the set of terms of $\mathcal{L}$.

5   Describe the set of terms of the first order language whose alphabet of symbols contains no individual contants and only the one function symbol f_1^1.

6   Which of the following are well-formed formulas?
(a) $A_1^2(f_1^1(x_1), x_1)$.
(b) $f_1^3(x_1, x_3, x_4)$.
(c) $(A_1^1(x_2) \rightarrow A_1^3(x_3, a_1))$.
(d) $\sim(\forall x_2)A_1^2(x_1, x_2)$.
(e) $((\forall x_2)A_1^1(x_1) \rightarrow (\sim A_1^1(x_2)))$.
(f) $A_1^3(f_2^3(x_1, x_2, x_3))$.
(g) $(\sim A_1^1(x_1) \rightarrow A_1^1(x_2))$.
(h) $(\forall x_1)A_1^3(a_1, a_2, f_1^1(a_3))$.

7   Which occurrences of x_1 in the following *wfs.* are free and which are bound?
(a) $(\forall x_2)(A_1^2(x_1, x_2) \rightarrow A_1^2(x_2, a_1))$.
(b) $(A_1^1(x_3) \rightarrow (\sim(\forall x_1)(\forall x_2)A_1^3(x_1, x_2, a_1)))$.
(c) $((\forall x_1)A_1^1(x_1) \rightarrow (\forall x_2)A_1^2(x_1, x_2))$.
(d) $(\forall x_2)(A_1^2(f_1^1(x_1, x_2), x_1) \rightarrow (\forall x_1)A_2^2(x_3, f_2^2(x_1, x_2)))$.
Is the term $f_1^2(x_1, x_3)$ free for x_2 in any or all of these?

8   Let $\mathcal{A}(x_i)$ be a *wf.* of $\mathcal{L}$ in which x_i occurs free, and let x_j be a variable which does not occur free in $\mathcal{A}(x_i)$. Show that if x_j is free for x_i in $\mathcal{A}(x_i)$ then x_i is free for x_j in $\mathcal{A}(x_j)$. $\mathcal{A}(x_j)$ is the result of substituting x_j for every free occurrence of x_i in $\mathcal{A}(x_i)$.

9   In each case below, let $\mathcal{A}(x_1)$ be the given *wf.*, and let t be the term $f_1^2(x_1, x_3)$. Write out the *wf.* $\mathcal{A}(t)$ and hence decide in each case whether t is free for x_1 in the given *wf.*
(a) $((\forall x_2)A_1^2(x_2, f_1^2(x_1, x_2)) \rightarrow A_1^1(x_1))$.
(b) $(\forall x_1)(\forall x_3)(A_1^1(x_3) \rightarrow A_1^1(x_1))$.
(c) $(\forall x_2)A_1^1(f_1^1(x_2)) \rightarrow (\forall x_3)A_1^3(x_1, x_2, x_3)$.
(d) $(\forall x_2)A_1^3(x_1, f_1^1(x_1), x_2) \rightarrow (\forall x_3)A_1^1(f_1^2(x_1, x_3))$.

10   Repeat Exercise 6 for the following terms t.
(a) x_2.
(b) x_3.
(c) $f_1^2(a_1, x_1)$.
(d) $f_1^3(x_1, x_2, x_3)$.

3.3 Interpretations

Definition 3.14

An *interpretation* I of $\mathscr{L}$ is a non-empty set D_I (the *domain* of I) together with a collection of distinguished elements $(\bar{a}_1, \bar{a}_2, \ldots)$, a collection of functions on D_I ($\bar{f}_i^n$, $i > 0$, $n > 0$), and a collection of relations on D_I ($\bar{A}_i^n$, $i > 0$, $n > 0$).

▷ It was pointed out before that the variables $x_1, x_2, \ldots$ of $\mathscr{L}$ were to be interpreted as 'objects'. The set D_I is to be the domain of objects over which the variables are supposed to range. The collection of distinguished elements is to consist of particular objects for which the individual constants of $\mathscr{L}$ stand. Likewise the relations and functions are to be concrete interpretations for the predicate letters and function letters of $\mathscr{L}$. Note that for a particular language $\mathscr{L}$, the lists of constants, function letters and predicate letters may be restricted. An interpretation of such a language $\mathscr{L}$ will have only sufficient distinguished elements, functions and relations to interpret the symbols appearing in $\mathscr{L}$.

We are now in a position to shed more light on the term 'first order language'. Such a language contains variables interpretable as objects in the domain of an interpretation, and quantifiers relating to these variables. In an interpretation, therefore, the quantifiers range over objects in the domain. This is the characteristic property of a first order language. A second order language would in addition contain quantifiers which are interpretable as ranging over relations among (and therefore sets of) objects in the domain of an interpretation. Such a language would have two sorts of variables, one sort for objects, another sort for relations. We shall restrict our attention entirely to first order languages and formal systems.

Example 3.15

Formal Arithmetic. See Example 3.5 (a) for a description of the appropriate first order language. It contains a_1, A_1^2, f_1^1, f_1^2, f_2^2, as well as variables, punctuation, connectives and quantifier. We may define an interpretation N as follows. Let $D_N = \{0, 1, 2, \ldots\}$, the set of natural numbers. The single distinguished element is 0 (the interpretation of the individual constant a_1). Addition and multiplication of natural numbers are the interpretations of the two two-place function letters f_1^2 and f_2^2 respectively, and the successor function is the interpretation of f_1^1. The relation $=$ is the interpretation of the predicate letter A_1^2.

Well-formed formulas in this language can, by means of the above, be interpreted as statements about natural numbers. For example, the *wf.*

$$(\forall x_1)(\forall x_2) \sim (\forall x_3)(\sim A_1^2(f_1^2(x_1, x_3), x_2)) \qquad (*)$$

has the interpretation ·

> for all x and $y \in D_N$, it is not the case that for every
> $z \in D_N, x + z \neq y$,

or equivalently

> for all x and $y \in D_N$, there is $z \in D_N$ such that
> $x + z = y$.

▷ An example of a statement about natural numbers which is not the interpretation of some *wf.* of the appropriate first order language is: 'every non-empty set of natural numbers has a least member'. This statement begins with a universal quantifier ranging over sets of numbers, and so would correspond to a *wf.* of a *second* order formal language for arithmetic.

It is only when an interpretation of the symbols of $\mathscr{L}$ is given that we can say anything about *meanings* of *wfs.*, and consequently it is only in the context of an interpretation that we can consider truth or falsity. The *wf.* (*) above turns out to have a meaning which is false in this interpretation, but in a different interpretation its meaning might well be true.

Example 3.16

The *wf.* (*) has a meaning which is true in the interpretation I, where D_I is the set of positive rational numbers, 1 is the interpretation of a_1, multiplication is the interpretation of f_1^2 and division (quotient function) is the interpretation of f_2^2.

(*) translates in I to

> for all $x, y \in D_I$ there is $z \in D_I$ such that $xz = y$.

This is a well known property of rational numbers.

▷ We include this last example principally to illustrate that there can be interpretations of the same formal language $\mathscr{L}$ which differ substantially. In our examples of languages we shall normally have some particular interpretation in mind, but this should not distract us from consideration of other interpretations and the consequences of their existence.

We may now see a similarity between this situation and the situation of Chapter 2. There the *wfs*. of the system L of statement calculus could not in themselves be regarded as true or false. Truth and falsity were relevant only when we assigned truth values to the statement variables or constructed a valuation. Furthermore, we saw that some *wfs*. could be either true or false depending on the valuation. In our present system, which is more complicated, the notion of interpretation will correspond to the assignment of truth values. An obvious question now to ask is: do we have an analogue of a tautology? The answer is 'yes', and the definition is just as one would expect, but we shall return later to consider it. Now we must make more precise the idea of *truth* in an interpretation. In spite of the apparent complexity of what follows, the ideas are not difficult, and the reader should keep in mind the intuitive picture that we have already built up. For the remainder of this chapter, it is suggested that the reader who is new to the subject should omit the details of the proofs on the first reading. The proofs are not important for an intuitive grasp of the ideas involved, and those ideas could be obscured by too careful attention to detail.

Exercises

11   Let $\mathscr{L}$ be the first order language which includes (besides variables, punctuation, connectives and quantifier) the individual constant a_1, the function letter f_1^2 and the predicate letter A_2^2. Let $\mathscr{A}$ denote the *wf*.

$$(\forall x_1)(\forall x_2)(A_2^2(f_1^2(x_1, x_2), a_1) \rightarrow A_2^2(x_1, x_2)).$$

Define an interpretation I of $\mathscr{L}$ as follows. D_I is $\mathbb{Z}$, $\bar{a}_1$ is 0, $\bar{f}_1^2(x, y)$ is $x - y$, $\bar{A}_2^2(x, y)$ is $x < y$. Write down the interpretation of $\mathscr{A}$ in I. Is this a true statement or a false one? Find another interpretation in which $\mathscr{A}$ is interpreted by a statement with the opposite truth value.

12   Is there an interpretation (of an appropriate language $\mathscr{L}$) in which the *wf*.

$$(\forall x_1)(A_1^1(x_1) \rightarrow A_1^1(f_1^1(x_1)))$$

is interpreted by a statement which is false? If so, describe one in detail. If not, explain why not.

13   Repeat Exercise 12 with the *wf*.

$$(\forall x_1)(A_1^2(x_1, x_2) \rightarrow A_1^2(x_2, x_1)).$$

3.4 Satisfaction, truth

Let I be an interpretation of the language $\mathscr{L}$, with domain D_I. Here and henceforward we shall use the notation in Definition 3.14. The interpretations in I of a_i, f_i^n, A_i^n respectively will be denoted by $\bar{a}_i$, $\bar{f}_i^n$, $\bar{A}_i^n$. Note that, for each i, $\bar{a}_i \in D_I$, $\bar{f}_i^n : D_I^n \rightarrow D_I$, and $\bar{A}_i^n$ is a n-place relation on D_I.

Definition 3.17

A *valuation* in I is a function v from the set of terms of $\mathscr{L}$ to the set D_I, with the properties:

(i) $v(a_i) = \bar{a}_i$ for each individual constant a_i of $\mathscr{L}$.

(ii) $v(f_i^n(t_1, \ldots, t_n)) = \bar{f}_i^n(v(t_1), \ldots, v(t_n))$, where f_i^n is any function letter in $\mathscr{L}$, and $t_1, \ldots, t_n$ are any terms of $\mathscr{L}$.

A valuation is thus just a rule which assigns to each term in $\mathscr{L}$ the object in D_I which is to be its interpretation. Part (ii) above ensures that the rule is a consistent one.

Remarks 3.18

(*a*) In general, in a given interpretation there will be many different valuations.

(*b*) A given valuation will assign an element of D_I to each of the variables x_i of $\mathscr{L}$. A valuation v will be completely specified by giving $v(x_1), v(x_2), \ldots$ This is because the $v(a_i)$ are given by the definition (i), and, inductively, for any term $f_i^n(t_1, \ldots, t_n)$, the value of $v(f_i^n(t_1, \ldots, t_n))$ is determined by (ii).

▷ We shall have occasion later to deal with valuations which are very nearly identical, in the following sense.

Definition 3.19

Two valuations v and v' are *i-equivalent* if $v(x_j) = v'(x_j)$ for every $j \neq i$.

Valuations which are *i*-equivalent have the same values on each of the variables, except possibly on x_i, but note that in general the values will differ on any term t in which x_i occurs.

▷ Now let us consider a *wf.* $\mathscr{A}$ of $\mathscr{L}$. A given valuation may have the effect of 'assigning a truth value' to $\mathscr{A}$ in the following way. Replace each term t appearing in $\mathscr{A}$ by $v(t)$, and replace each function letter and predicate letter by its interpretation in I. What is obtained is a statement about the elements of the set D_I which may be true and may be false. If it is true we say that v satisfies $\mathscr{A}$. We make this precise as follows.

Definition 3.20

Let $\mathscr{A}$ be a *wf.* of $\mathscr{L}$, and let I be an interpretation of $\mathscr{L}$. A valuation v in I is said to *satisfy* $\mathscr{A}$ if it can be shown inductively to do so under the following four conditions.

(i) v satisfies the atomic formula $A_j^n(t_1, \ldots, t_n)$ if $\bar{A}_j^n(v(t_1), \ldots, v(t_n))$ is true in D_I.

(ii) v satisfies $(\sim\mathscr{B})$ if v does not satisfy $\mathscr{B}$.

(iii) v satisfies $(\mathscr{B} \rightarrow \mathscr{C})$ if either v satisfies $(\sim\mathscr{B})$ or v satisfies $\mathscr{C}$.

(iv) v satisfies $(\forall x_i)\mathscr{B}$ if every valuation v' which is i-equivalent to v satisfies $\mathscr{B}$.

Remarks 3.21

(a) For any v and $\mathscr{A}$, either v satisfies $\mathscr{A}$ or v satisfies $(\sim\mathscr{A})$.

(b) Some explanantion of (iv) is perhaps necessary. $(\forall x_i)\mathscr{B}$ will be interpreted by some statement 'for every $y \in D_I, \ldots$', y being thought of as the interpretation of x_i. v provides interpretations for the variables which appear in $\mathscr{B}$, and it is reasonable to say that v satisfies $(\forall x_i)\mathscr{B}$ if first of all v satisfies $\mathscr{B}$ and further any valuation which is obtained from v by changing $v(x_i)$ also satisfies $\mathscr{B}$.

(c) It may aid understanding of this definition to point out that each of the clauses (i) to (iv) above may be regarded as 'if and only if' statements, since each is in the nature of a definition.

Example 3.22

(a) In the arithmetic interpretation N, consider the wf. $A_1^2(f_2^2(x_1, x_2), f_2^2(x_3, x_4))$. Any valuation v in which $v(x_1) = 2$, $v(x_2) = 6$, $v(x_3) = 3$, $v(x_4) = 4$ will satisfy this wf., since $2 \times 6 = 3 \times 4$ is true in D_N. Likewise, any valuation w with $w(x_1) = 1$, $w(x_2) = 5$, $w(x_3) = 4$, $w(x_4) = 2$ will not satisfy it.

(b) The wf. $(\forall x_1)A_1^2(f_2^2(x_1, x_2), f_1^2(x_2, x_1))$ is interpreted in N as 'for every $n \in D_N$, $nm = mn$', which would certainly be regarded as true. Let v be a valuation in N. Then $v(x_1)$ and $v(x_2)$ are natural numbers, and $A_1^2(f_2^2(x_1, x_2), f_2^2(x_2, x_1))$ is then interpreted as $v(x_1) \times v(x_2) = v(x_2) \times v(x_1)$, which is certainly true. So v satisfies $A_1^2(f_2^2(x_1, x_2), f_2^2(x_2, x_1))$. Moreover, if we change the value of $v(x_1)$ to get a new 1-equivalent valuation v', it is easy to see that v' also satisfies $A_1^2(f_2^2(x_1, x_2), f_2^2(x_2, x_1))$. Hence, by (iv) of the definition above, v satisfies $(\forall x_1)A_1^2(f_2^2(x_1, x_2), f_2^2(x_2, x_1))$ in N. Thus every valuation v in N satisfies this wf.

(c) The wf. $(\forall x_1)A_1^2(x_1, a_1)$ is interpreted intuitively as 'for every $n \in D_N$, $n = 0$', which is false. Let v be a valuation in N. Then $A_1^2(x_1, a_1)$ is interpreted by '$v(x_1) = 0$'. So v does not satisfy $A_1^2(x_1, a_1)$ unless $v(x_1) = 0$. Suppose then that $v(x_1) = 0$, and ask whether v satisfies $(\forall x_1)A_1^2(x_1, a_1)$. We must be able to say that any v' obtained from v by changing the value of $v(x_1)$ also satisfies $A_1^2(x_1, a_1)$. We clearly cannot. Hence no valuation in N satisfies $(\forall x_1)A_1^2(x_1, a_1)$.

▷ Substitution for variables by other variables or terms is an important technique later. The following proposition will be necessary, though its technical nature makes the proof complicated. It illustrates the form that proofs take in this part of the subject.

Proposition 3.23

Let $\mathcal{A}(x_i)$ be a *wf.* of $\mathcal{L}$ in which x_i appears free, and let t be a term which is free for x_i in $\mathcal{A}(x_i)$. Suppose that v is a valuation and v' is the valuation which is i-equivalent to v and has $v'(x_i) = v(t)$. Then v satisfies $\mathcal{A}(t)$ if and only if v' satisfies $\mathcal{A}(x_i)$.

Proof. First observe that for any term u in which x_i occurs we can obtain a term u' by substituting t for x_i throughout, and that then $v(u') = v'(u)$. We prove this by induction on the length of u (i.e. the number of symbols in u).

Base step: u is x_i, so u' is t. Then

$$v'(u) = v'(x_i) = v(t), \text{ (by definition of } v')$$

$$= v(u').$$

Induction step: u is $f_i^n(u_1, \ldots, u_n)$, where $u_1, \ldots, u_n$ are terms with smaller length. Let $u_1', \ldots, u_n'$ be obtained by substituting t for x_i throughout. Then $u' = f_i^n(u_1', \ldots, u_n')$. So

$$v(u') = \bar{f}_i^n(v(u_1'), \ldots, v(u_n'))$$

$$= \bar{f}_i^n(v'(u_1), \ldots, v'(u_n)),$$

by the induction hypothesis,

$$= v'(u).$$

Thus $v(u') = v'(u)$ has been established, for every term u of $\mathcal{L}$.

We prove the proposition now by induction on the length of the *wf.* $\mathcal{A}(x_i)$, i.e. the number of connectives and quantifiers in $\mathcal{A}(x_i)$.

Base step: $\mathcal{A}(x_i)$ is an atomic formula, say $A_j^n(u_1, \ldots, u_n)$ where $u_1, \ldots, u_n$ are terms in $\mathcal{L}$. Suppose that v' satisfies $\mathcal{A}(x_i)$. Then

$$\bar{A}_j^n(v'(u_1), \ldots, v'(u_n)) \text{ holds in the interpretation,}$$

so

$$\bar{A}_j^n(v(u_1'), \ldots, v(u_n')) \text{ holds in the interpretation,}$$

where $u_1', \ldots, u_n'$ are obtained as above by substituting t for x_i throughout. (Here we use the preliminary result above.) We now have that v satisfies the *wf.* $A_j^n(u_1', \ldots, u_n')$, i.e. v satisfies $A(t)$. The converse can be demonstrated by retracing this argument.

Induction step:

Case 1: $\mathcal{A}(x_i)$ is $\mathcal{B}(x_i)$.

Case 2: $\mathcal{A}(x_i)$ is $(\mathcal{B}(x_i) \rightarrow \mathcal{C}(x_i))$.

These are straightforward and left as exercises.

Case 3: $\mathcal{A}(x_i)$ is $(\forall x_j)\mathcal{B}(x_i)$ $(j \neq i)$.

Suppose that v does not satisfy $\mathcal{A}(t)$. We show that v' does not satisfy $\mathcal{A}(x_i)$. There is a valuation w which is j-equivalent to v and which does not satisfy $\mathcal{B}(t)$. Let w' be the valuation which is i-equivalent to w, having $w'(x_i) = w(t)$. Then by the induction hypothesis applied to $\mathcal{B}(x_i)$, we have that w' does not satisfy $\mathcal{B}(x_i)$ (since w does not satisfy $\mathcal{B}(t)$). Now t is free for x_i in $(\forall x_j)\mathcal{B}(x_i)$, so x_j does not occur in t. Hence $v(t)$ depends only on $v(x_k)$ for $k \neq j$. But for $k \neq j$, $v(x_k) = w(x_k)$, so $v(t) = w(t)$. It follows that w' is j-equivalent to v', since w is j-equivalent to v. Since w' does not satisfy $\mathcal{B}(x_i)$, then, v' does not satisfy $(\forall x_j)\mathcal{B}(x_i)$, i.e. v' does not satisfy $\mathcal{A}(x_i)$. The converse requires a similar argument, and is left as an exercise.

Definition 3.24

A *wf.* $\mathcal{A}$ is *true* in an interpretation I if every valuation in I satisfies $\mathcal{A}$. $\mathcal{A}$ is *false* if there is no valuation in I which satisfies $\mathcal{A}$.

Notation. We write $I \vDash \mathcal{A}$ if $\mathcal{A}$ is true in I. This symbol should not be confused with $\vdash$, but the reader should note that neither is a symbol in the formal language. Each of them is part of the metalanguage which we use to talk about our formal systems.

Remarks 3.25

(*a*) It may happen that, for a particular *wf.* $\mathcal{A}$, some valuations in I satisfy $\mathcal{A}$ and some valuations do not. Such a *wf.* is neither true nor false in I.

(*b*) The domain of an interpretation is by definition non-empty, so trivially the set of valuations cannot be empty. It is clear from the definition that a given valuation either satisfies or does not satisfy a given *wf.* $\mathcal{A}$, and hence it is impossible for a *wf.* to be both true and false in a given interpretation.

(*c*) In a given interpretation, a *wf.* $\mathcal{A}$ is false if and only if $(\sim\mathcal{A})$ is true. This is an immediate consequence of the definitions of valuation and truth. It follows that for no *wf.* $\mathcal{A}$ can both $\mathcal{A}$ and $(\sim\mathcal{A})$ be true.

(*d*) In a given interpretation I, a *wf.* $(\mathcal{A} \to \mathcal{B})$ is false if and only if $\mathcal{A}$ is true and $\mathcal{B}$ is false. Let us prove one direction, to illustrate how the definitions are applied. Suppose that $(\mathcal{A} \to \mathcal{B})$ is false in I. Then no valuation satisfies $(\mathcal{A} \to \mathcal{B})$ in I. Given any valuation v, then, v does not satisfy $(\mathcal{A} \to \mathcal{B})$. By Definition 3.20 (iii), v does not satisfy $(\sim\mathcal{A})$ and v does not satisfy $\mathcal{B}$. Hence v satisfies $\mathcal{A}$ and v does not satisfy $\mathcal{B}$. Thus every valuation satisfies $\mathcal{A}$ and does not satisfy $\mathcal{B}$. It follows that $\mathcal{A}$ is true and $\mathcal{B}$ is false.

Proposition 3.26

If, in a particular interpretation I, the *wfs.* $\mathscr{A}$ and $(\mathscr{A} \to \mathscr{B})$ are true, then $\mathscr{B}$ is also true.

Proof. Let v be a valuation in I. Then v satisfies $\mathscr{A}$ and v satisfies $(\mathscr{A} \to \mathscr{B})$. By the definition of satisfaction, then, either v satisfies $(\sim\mathscr{A})$ or v satisfies $\mathscr{B}$. But v cannot satisfy $(\sim\mathscr{A})$, so v must satisfy $\mathscr{B}$. It follows that $\mathscr{B}$ is satisfied by every valuation in I, so $\mathscr{B}$ is true in I.

Proposition 3.27

Let $\mathscr{A}$ be a *wf.* of $\mathscr{L}$, and let I be an interpretation of $\mathscr{L}$. Then $I \vDash \mathscr{A}$ if and only if $I \vDash (\forall x_i)\mathscr{A}$, where x_i is any variable.

Proof. Suppose that $I \vDash \mathscr{A}$, and let v be any valuation in I. Then v satisfies $\mathscr{A}$ and every v' which is i-equivalent to v also satisfies $\mathscr{A}$, since every valuation satisfies $\mathscr{A}$. Hence v satisfies $(\forall x_i)\mathscr{A}$, and so every valuation satisfies $(\forall x_i)\mathscr{A}$, i.e. $I \vDash (\forall x_i)\mathscr{A}$.

Now suppose that $I \vDash (\forall x_i)\mathscr{A}$, and let v be any valuation in I. Then v satisfies $(\forall x_i)\mathscr{A}$. Hence every v' which is i-equivalent to v satisfies $\mathscr{A}$. In particular, v satisfies $\mathscr{A}$, and so every valuation satisfies $\mathscr{A}$, i.e. $I \vDash \mathscr{A}$.

Corollary 3.28

Let $y_1, \ldots, y_n$ be variables in $\mathscr{L}$, let $\mathscr{A}$ be a *wf.* of $\mathscr{L}$, and let I be an interpretation. Then $I \vDash \mathscr{A}$ if and only if $I \vDash (\forall y_1) \ldots (\forall y_n)\mathscr{A}$.

Proof. By repeated application of Proposition 3.27.

▷ There are two aspects of this result which are worth mentioning. Firstly, adding a quantifier for a variable not occurring free in $\mathscr{A}$ to get $(\forall x_i)\mathscr{A}$ intuitively does not change the interpretation (see Example 3.10), so it should not be surprising that in this circumstance $\mathscr{A}$ should be true if and only if $(\forall x_i)\mathscr{A}$ is true. Adding a quantifier for a variable which does appear free in $\mathscr{A}$, to get $(\forall x_i)\mathscr{A}(x_i)$, has a different effect. But the result above indicates that $\mathscr{A}(x_i)$ is true if and only if $(\forall x_i)\mathscr{A}(x_i)$ is true, so when we are considering the truth or falsity of *wfs.* with free variables there is a sense in which the universal quantifier(s) is (are) understood.

The existential quantifier $\exists$ has been introduced as a defined symbol into the formal language. Let us see how this fits in the context of valuations and satisfaction.

Proposition 3.29

In an interpretation I, a valuation v satisfies the formula $(\exists x_i)\mathscr{A}$ if and only if there is at least one valuation v' which is i-equivalent to v and which satisfies $\mathscr{A}$.

Proof. $(\exists x_i)\mathscr{A}$ stands for $\sim(\forall x_i)(\sim\mathscr{A})$. Let v satisfy $\sim(\forall x_i)(\sim\mathscr{A})$. Then v does not satisfy $(\forall x_i)(\sim\mathscr{A})$. Thus there is some v' which is i-equivalent to v which does not satisfy $(\sim\mathscr{A})$. This v' must then satisfy $\mathscr{A}$. The converse is proved by reversing the above argument.

▷ The language of propositional calculus contained the connectives $\sim$ and $\to$. So also does the language $\mathscr{L}$. Thus if we take a *wf.* $\mathscr{A}_0$ of L and replace each statement letter occurring by a *wf.* of $\mathscr{L}$ (replacing the same letter by the same *wf.* throughout), we shall obtain a *wf.* $\mathscr{A}$ of $\mathscr{L}$. $\mathscr{A}$ is then called a *substitution instance* in $\mathscr{L}$ of $\mathscr{A}_0$. Similarly, starting with a *wf.* of $\mathscr{L}$, we can see that it will have the same structure as some (usually more than one) *wf.* of L. For example, consider

$$(\sim(\forall x_1)A_1^1(x_1) \to ((\forall x_2)A_2^2(x_1, x_2) \to (A_1^1(x_2) \to (\forall x_1)A_3^1(x_1)))).$$

This is a substitution instance of the *wf.* $(\sim p_1 \to p_2)$ of L, and also of $(\sim p_1 \to (p_2 \to p_3))$. As another example, consider

$$((\forall x_1)A_1^1(x_1) \to (\forall x_1)A_1^1(x_1)).$$

This *wf.* of $\mathscr{L}$ is a substitution instance of the *wf.* $(p_1 \to p_1)$ of L. Now $(p_1 \to p_1)$ is a tautology. We extend the notion of tautology to *wfs.* of $\mathscr{L}$ as follows.

Definition 3.30

A *wf.* $\mathscr{A}$ of $\mathscr{L}$ is a *tautology* if it is a substitution instance in $\mathscr{L}$ of a tautology in L.

Proposition 3.31

A *wf.* of $\mathscr{L}$ which is a tautology is true in any interpretation of $\mathscr{L}$.

Proof. It is the analogy between Definitions 2.12 and 3.20 which yields our method. Let $\mathscr{A}$ be a *wf.* of $\mathscr{L}$ which is a substitution instance of a *wf.* $\mathscr{A}_0$ of L. From any valuation v in an interpretation I we can obtain a (partial) valuation v' of L as follows. Let $p_1, \ldots, p_k$ be the statement letters occurring in $\mathscr{A}_0$, and let $\mathscr{A}_1, \ldots, \mathscr{A}_k$ be the *wfs.* of $\mathscr{L}$ which are substituted for them respectively to obtain $\mathscr{A}$. For $1 \le i \le k$, let

$$v'(p_i) = \begin{cases} T & \text{if } v \text{ satisfies } \mathscr{A}_i \\ F & \text{if } v \text{ does not satisfy } \mathscr{A}_i. \end{cases}$$

We now show that v satisfies $\mathscr{A}$ if and only if $v'(\mathscr{A}_0) = T$. The proof is by induction on the number of connectives in $\mathscr{A}_0$.

Base step: $\mathscr{A}_0$ is a statement letter, p_n, say. By the definition of v', then, $v'(p_n) = T$ if and only if v satisfies $\mathscr{A}$.

Induction step:

Case 1: $\mathcal{A}_0$ is $\sim\mathcal{B}_0$, and so $\mathcal{A}$ is $\sim\mathcal{B}$, say, where $\mathcal{B}$ is a substitution instance of $\mathcal{B}_0$. By the induction hypothesis, v satisfies $\mathcal{B}$ if and only if $v'(\mathcal{B}_0) = T$. Hence v does not satisfy $\mathcal{B}$ if and only if $v'(\mathcal{B}_0) = F$. It follows that v satisfies $\mathcal{A}$ if and only if $v'(\mathcal{A}_0) = T$, using Definitions 3.20 (ii) and 2.12 (i).

Case 2: $\mathcal{A}_0$ is $(\mathcal{B}_0 \to \mathcal{C}_0)$, and so $\mathcal{A}$ is $(\mathcal{B} \to \mathcal{C})$, where $\mathcal{B}$ and $\mathcal{C}$ are substitution instances of $\mathcal{B}_0$ and $\mathcal{C}_0$. The following assertions are all equivalent:

(a) v satisfies $\mathcal{A}$.
(b)   either v satisfies $\sim\mathcal{B}$ or v satisfies $\mathcal{C}$ (by Definition 3.20 (iii)).
(c)   either v does not satisfy $\mathcal{B}$ or v satisfies $\mathcal{C}$.
(d)   either $v'(\mathcal{B}_0) = F$ or $v'(\mathcal{C}_0) = T$.
(e) $v'(\mathcal{B}_0 \to \mathcal{C}_0) = T$ (by Definition 2.12).
(f) $v'(\mathcal{A}_0) = T$.

This concludes the inductive proof. The proof of the proposition is now straightforward.

Let $\mathcal{A}$ be a *wf.* of $\mathcal{L}$ which is a tautology. Then $\mathcal{A}$ is a substitution instance of a tautology $\mathcal{A}_0$ in L. Let v be a valuation in an interpretation I. From the above we see that v satisfies $\mathcal{A}$ if $v'(\mathcal{A}_0) = T$. But $\mathcal{A}_0$ is a tautology, so v satisfies $\mathcal{A}$. Thus $\mathcal{A}$ is true in I.

▷ We have seen that in a given interpretation not every *wf.* need be true or false. Consider for example the *wf.* $A_1^1(x_1)$, and an interpretation I in which D_I is $\mathbb{Z}$, the set of integers, and the interpretation of A_1^1 is the predicate '>0'. Then $A_1^1(x_1)$ is satisfied by any valuation v for which $v(x_1) > 0$, but $A_1^1(x_1)$ is satisfied by no valuation w for which $w(x_1) \leq 0$. Intuitively, whether $A_1^1(x_1)$ is made true or not depends on the interpretation of x_1. This situation arises very often when we are dealing with a *wf.* in which there is an occurrence of a free variable. Our next main result says that a *wf.* in which there is *no* occurrence of a free variable must be either true or false in a given interpretation. We need some preliminary work first.

Definition 3.32

A *wf.* $\mathcal{A}$ of $\mathcal{L}$ is said to be *closed* if no variable occurs free in $\mathcal{A}$.

Proposition 3.33

Let I be an interpretation of $\mathcal{L}$ and let $\mathcal{A}$ be a *wf.* of $\mathcal{L}$. If v and w are valuations such that $v(x_i) = w(x_i)$ for every free variable x_i of $\mathcal{A}$, then v satisfies $\mathcal{A}$ if and only if w satisfies $\mathcal{A}$.

Proof. By induction on the number of connectives and quantifiers in $\mathcal{A}$.

Base step: $\mathscr{A}$ is an atomic formula, say $A_i^n(t_1, \ldots, t_n)$. The valuations v and w agree on the free variables which occur in $t_1, \ldots, t_n$, and on any individual constants which appear, so $v(t_i) = w(t_i)$ for $1 \le i \le n$. Therefore v satisfies $\mathscr{A}$ if and only if w satisfies $\mathscr{A}$.

Induction step:

Case 1: $\mathscr{A}$ is $\sim\mathscr{B}$.

Case 2: $\mathscr{A}$ is $(\mathscr{B} \to \mathscr{C})$.

These two cases are straightforward, using merely the appropriate definitions. They are left as exercises.

Case 3: $\mathscr{A}$ is $(\forall x_i)\mathscr{B}$.

Suppose that v satisfies $\mathscr{A}$, and let w' be i-equivalent to w. Then since x_i does not occur free in $(\forall x_i)\mathscr{B}$, we have $v(y) = w'(y)$ whenever y is a free variable of $\mathscr{A}$. Now every v' which is i-equivalent to v satisfies $\mathscr{B}$, so in particular let v' be determined by

$$v'(x_i) = w'(x_i)$$

$$v'(x_j) = v(x_j) \quad \text{if } j \ne i.$$

Then $w'(y) = v'(y)$ whenever y is a free variable of $\mathscr{B}$. Hence by the induction hypothesis, since v' satisfies $\mathscr{B}$, we have w' satisfies $\mathscr{B}$, and so w satisfies $(\forall x_i)\mathscr{B}$.

We prove that if w satisfies $(\forall x_i)\mathscr{B}$ then v satisfies $(\forall x_i)\mathscr{B}$ in precisely the same way.

This completes the induction and the result is proved.

Corollary 3.34

If $\mathscr{A}$ is a closed *wf.* of $\mathscr{L}$ and I is an interpretation of $\mathscr{L}$, then either $I \vDash \mathscr{A}$ or $I \vDash (\sim\mathscr{A})$.

Proof. Let $\mathscr{A}$ be a closed *wf.* and let I be an interpretation. Let v and w be any valuations. Trivially, $v(y) = w(y)$ whenever y is a free variable of $\mathscr{A}$ ($\mathscr{A}$ has no free variables), so v satisfies $\mathscr{A}$ if and only if w satisfies $\mathscr{A}$. Hence either every valuation satisfies $\mathscr{A}$ or no valuation satisfies $\mathscr{A}$, i.e. either $\mathscr{A}$ is true in I or $\mathscr{A}$ is false in I. By Remark 3.25 (c), then, either $I \vDash \mathscr{A}$ or $I \vDash (\sim\mathscr{A})$.

$\triangleright$ This is a significant result for us, because the notion of truth in an interpretation is a more important one than that of satisfaction by a valuation. What we have established is that for *closed* formulas all valuations in a particular interpretation give us the same answer regarding satisfaction of a given formula. Hence to check on truth or falsity of a given closed *wf.* we need check only whether some valuation satisfies it or not. We shall see also that closed formulas are more relevant where mathematics itself is concerned – indeed formulas with free variables sometimes behave somewhat unnaturally.

Interpretations give truth values to the closed *wfs.* of $\mathscr{L}$. It may happen that for a given *wf.* $\mathscr{A}$ of $\mathscr{L}$ all interpretations of $\mathscr{L}$ give it the value T, i.e. $\mathscr{A}$ is true in all possible interpretations of $\mathscr{L}$.

Definition 3.35

A *wf.* $\mathscr{A}$ of $\mathscr{L}$ is *logically valid* if $\mathscr{A}$ is true in every interpretation of $\mathscr{L}$. $\mathscr{A}$ is *contradictory* if it is false in every interpretation.

These notions are the analogues in the present situation of the notions of tautology and contradiction from Chapter 1. Just as then there were some formulas which were neither tautology nor contradiction so in $\mathscr{L}$ there are some *wfs.* which are neither logically valid nor contradictory.

Remark 3.36

(*a*) It is an immediate consequence of Proposition 3.26 that if the *wfs.* $\mathscr{A}$ and $(\mathscr{A} \to \mathscr{B})$ are logically valid, then $\mathscr{B}$ is logically valid.

(*b*) In like manner, it is a consequence of Proposition 3.27 that if the *wf.* $\mathscr{A}$ is logically valid, then so is $(\forall x_i)\mathscr{A}$, where x_i is any variable.

Example 3.37

(*a*) We have seen that every substitution instance in $\mathscr{L}$ of a tautology in L is logically valid (Proposition 3.31). Note therefore that the class of logically valid *wfs.* of $\mathscr{L}$ contains the class of tautologies.

(*b*) $((\forall x_i)\mathscr{A} \to (\exists x_i)\mathscr{A})$ is logically valid, for any *wf.* $\mathscr{A}$ of $\mathscr{L}$. This is proved by a standard method as follows.

Let I be an interpretation with domain D_I and let v be a valuation in I. If v does not satisfy $(\forall x_i)\mathscr{A}$ then v satisfies $((\forall x_i)\mathscr{A} \to (\exists x_i)\mathscr{A})$. If v does satisfy $(\forall x_i)\mathscr{A}$ then every valuation v' which is i-equivalent to v satisfies $\mathscr{A}$. Clearly, then, there is one valuation which is i-equivalent to v which satisfies $\mathscr{A}$. So v satisfies $(\exists x_i)\mathscr{A}$, by Proposition 3.29. Hence in this case, also v satisfies $((\forall x_i)\mathscr{A} \to (\exists x_i)\mathscr{A})$. We have thus proved that an arbitrary valuation in an arbitrary interpretation satisfies the given *wf.* Hence this *wf.* is logically valid.

(*c*) $((\forall x_1)(\exists x_2)A_1^2(x_1, x_2) \to (\exists x_1)(\forall x_2)A_1^2(x_1, x_2))$ is not logically valid. The proof of this is perhaps slightly less straightforward, since what we have to do is find an interpretation in which the given *wf.* is not true. We have to choose a domain D_I, an interpretation for the predicate letter A_1^2, and a valuation which does not satisfy the *wf.*

Let $D_I = \mathbb{Z}$. Let $\bar{A}_1^2(y, z)$ be '$y < z$'. It is clear now without choosing a valuation that the closed *wf.* $(\forall x_1)(\exists x_2)A_1^2(x_1, x_2)$ is true in this interpretation, and $(\exists x_1)(\forall x_2)A_1^2(x_1, x_2)$ is false, i.e. every valuation satisfies the former and does not satisfy the latter. Hence no valuation satisfies the given *wf*. It is thus not true in this interpretation and it cannot be logically valid.

(*d*) $A_1^1(x_1)$ is not logically valid. As we saw before (p. 66), not only is this *wf*. not logically valid, but there are interpretations in which it is neither true nor false.

▷ In general, to prove logical validity we must show that an arbitrary valuation in an arbitrary interpretation satisfies the *wf*. concerned. To prove that a *wf*. is not logically valid usually requires some ingenuity in actually constructing an interpretation in which there is a valuation which does not satisfy it.

Exercises

14 In the arithmetic interpretation N of Example 3.5, find, if possible, valuations which satisfy and do not satisfy each of the following *wfs*.
   (*a*) $A_1^2(f_1^2(x_1, x_1), f_2^2(x_2, x_3))$.
   (*b*) $(A_1^2(f_1^2(x_1, a_1), x_2) \to A_1^2(f_1^2(x_1, x_2), x_3))$.
   (*c*) $\sim A_1^2(f_2^2(x_1, x_2), f_2^2(x_2, x_3))$.
   (*d*) $(\forall x_1)A_1^2(f_2^2(x_1, x_2), x_3)$.
   (*e*) $((\forall x_1)A_1^2(f_2^2(x_1, a_1), x_1) \to A_1^2(x_1, x_2))$.

15 In the interpretation described in Exercise 11, find, if possible, valuations which satisfy and do not satisfy each of the following *wfs*.
   (*a*) $A_2^2(x_1, a_1)$.
   (*b*) $A_2^2(f_1^2(x_1, x_2), x_1) \to A_2^2(a_1, f_1^2(x_1, x_2))$.
   (*c*) $\sim A_2^2(x_1, f_1^2(x_1, f_1^2(x_1, x_2)))$.
   (*d*) $(\forall x_1)A_2^2(f_1^2(x_1, x_2), x_3)$.
   (*e*) $(\forall x_1)A_2^2(f_1^2(x_1, a_1), x_1) \to A_2^2(x_1, x_2)$.

16 Which of the following closed *wfs*. are true in the interpretation N, and which are false?
   (*a*) $(\forall x_1)A_1^2(f_2^2(x_1, a_1), x_1)$.
   (*b*) $(\forall x_1)(\forall x_2)(A_1^2(f_1^2(x_1, a_1), x_2) \to A_1^2(f_1^2(x_2, a_1), x_1))$.
   (*c*) $(\forall x_1)(\forall x_2)(\exists x_3)A_1^2(f_1^2(x_1, x_2), x_3)$.
   (*d*) $(\exists x_1)A_1^2(f_1^2(x_1, x_1), f_2^2(x_1, x_1))$.

17 Which of the following closed *wfs*. are true in the interpretation of Exercise 11, and which are false?
   (*a*) $(\forall x_1)A_2^2(f_1^2(a_1, x_1), a_1)$.
   (*b*) $(\forall x_1)(\forall x_2)(\sim A_2^2(f_1^2(x_1, x_2), x_1))$.
   (*c*) $(\forall x_1)(\forall x_2)(\forall x_3)(A_2^2(x_1, x_2) \to A_2^2(f_1^2(x_1, x_3), f_1^2(x_2, x_3)))$.
   (*d*) $(\forall x_1)(\exists x_2)A_2^2(x_1, f_1^2(f_1^2(x_1, x_2), x_2))$.

18 Prove that, in a given interpretation, the *wf*. $(\mathscr{A} \to \mathscr{B})$ is false if and only if $\mathscr{A}$ is true and $\mathscr{B}$ is false. [See Remark 3.25 (*d*).]

19 Show that each of the following *wfs*. is logically valid.
   (*a*) $((\exists x_1)(\forall x_2)A_1^2(x_1, x_2) \to (\forall x_2)(\exists x_1)A_1^2(x_1, x_2))$.
   (*b*) $(\forall x_1)A_1^1(x_1) \to ((\forall x_1)A_2^1(x_1) \to (\forall x_2)A_1^1(x_2))$.

(c) $(\forall x_1)(\mathcal{A} \to \mathcal{B}) \to ((\forall x_1)\mathcal{A} \to (\forall x_1)\mathcal{B})$, for any *wfs.* $\mathcal{A}$ and $\mathcal{B}$.

(d) $((\forall x_1)(\forall x_2)\mathcal{A} \to (\forall x_2)(\forall x_1)\mathcal{A})$, for any *wf.* $\mathcal{A}$.

20   Give an example of a logically valid *wf.* which is not closed.

21   Show that, if t is a term which is free for x_i in the *wf.* $\mathcal{A}(x_i)$, then the *wf.* $(\mathcal{A}(t) \to (\exists x_i)\mathcal{A}(x_i))$ is a logically valid *wf.*

22   Show that none of the following *wfs.* is logically valid.

(a) $(\forall x_1)(\exists x_2)A_1^2(x_1, x_2) \to (\exists x_2)(\forall x_1)A_1^2(x_1, x_2)$.

(b) $\cdot (\forall x_1)(\forall x_2)(A_1^2(x_1, x_2) \to A_1^2(x_2, x_1))$.

(c) $\cdot (\forall x_1)((\sim A_1^1(x_1)) \to (\sim A_1^1(a_1)))$.

(d) $((\forall x_1)A_1^2(x_1, x_1) \to (\exists x_2)(\forall x_1)A_1^2(x_1, x_2))$.

23   Let $\mathcal{A}(x_i)$ be a *wf.* of $\mathcal{L}$ in which x_i appears free and let t be a term which is free for x_i in $\mathcal{A}(x_i)$. Suppose that v is a valuation such that $v(t) = v(x_i)$. Show that v satisfies $\mathcal{A}(t)$ if and only if v satisfies $\mathcal{A}(x_i)$.

3.5 Skolemisation

Consider the *wf.*

$$(\forall x_1)(\exists x_2)\mathcal{B}(x_1, x_2). \tag{*}$$

Let us suppose, for the sake of simplicity, that $\mathcal{B}(x_1, x_2)$ is a *wf.* in which only x_1 and x_2 occur free. We can think of this *wf.* as asserting:

> For all x_1 there is x_2 such that x_1 and x_2 are in the relationship expressed by $\mathcal{B}(x_1, x_2)$.

In other words, this describes a correspondence between values represented by x_1 and x_2. There need not be a unique value of x_2 associated with each value of x_1, but we can imagine a choice being made for each x_1, and thereby a function being defined. The sense of the *wf.*

$$(\forall x_1)\mathcal{B}(x_1, h_1^1(x_1))$$

is similar to the sense of the *wf.* (*), provided that the function letter h_1^1 is distinct from all other function letters occurring. This *wf.* is a *Skolemised form* of (*), and the function h_1^1 is called a *Skolem function*. To avoid all possibility of confusion, we take Skolem function symbols from a completely new collection of symbols, distinct from those of Section 3.2. For positive integers i and n, h_i^n will be an n-place Skolem function symbol.

A Skolemised form of a *wf.* certainly does not carry the same meaning, so we cannot expect the two *wfs.* to be logically equivalent. But there is a weak sense in which they are equivalent: if one is contradictory then the other is contradictory. But before considering this further, let us extend the notion of Skolemisation.

Suppose that the *wf.* $(\exists x_i)\mathcal{B}$ occurs as a subformula in a *wf.* $\mathcal{A}$, within the scopes of the universal quantifiers $(\forall x_{i_1}), \ldots, (\forall x_{i_r})$. Then we may write

$\mathscr{B}$ as $\mathscr{B}(x_{i_1}, ..., x_{i_r}, x_i)$, although not all of these variables need occur explicitly in $\mathscr{B}$. Now delete the existential quantifier $(\exists x_i)$ and replace all occurrences of x_i by $h_1^r(x_{i_1}, ..., x_{i_r})$ (say).

This describes one step in a process which can be used to eliminate all existential quantifiers from any given *wf*. One case not covered above, however, is when an existential quantifier occurs but not within the scope of a universal quantifier. Occurrences of the quantified variable would in this case be replaced by a constant symbol rather than a functional term. Thus besides Skolem function letters h_i^n, we shall need a collection of Skolem constant letters, say c_i.

Examples 3.38

(a) $(\forall x_1)(\exists x_2)(A_2^1(x_1, x_2) \rightarrow (\exists x_3) A_1^3(x_1, x_2, x_3))$

yields a Skolemised form

$$(\forall x_1)(A_1^2(x_1, h_1^1(x_1)) \rightarrow A_1^3(x_1, h_1^1(x_1), h_2^1(x_1))).$$

(b) $(\exists x_1)(\forall x_2)((A_1^1(x_1) \wedge (\forall x_3) A_1^3(x_1, x_2, x_3))$
        $\rightarrow (\exists x_4)(\forall x_5) A_1^3(x_1, x_4, x_5))$

yields a Skolemised form

$$(\forall x_2)((A_1^1(x_2) \wedge (\forall x_3) A_1^3(c_1, x_2, x_3))$$
$$\rightarrow (\forall x_5) A_1^3(c_1, h_1^1(x_2), x_5))$$

▷ The Skolemisation process gives a result which is independent of the order in which the existential quantifiers are removed, as can be seen easily in the above simple cases. Thus, apart from variations of the letters chosen as Skolem functions and constants, each *wf*. will have a unique Skolemised form.

Proposition 3.39

A *wf*. $\mathscr{A}$ of $\mathscr{L}$ is contradictory if and only if its Skolemised form is contradictory.

Proof. A proof of this in full generality is too substantial to be included here, but we can give the proof for a simple case, in order to give the flavour. Suppose that the *wf*. has the form $(\forall x_1)(\exists x_2) \mathscr{B}(x_1, x_2)$, so that $\mathscr{A}^S$, its Skolemised form, is $(\forall x_1) \mathscr{B}(x_1, h_1^1(x_1))$. It will be easier to prove the Proposition in a modified (but equivalent) form: there is an interpretation in which $\mathscr{A}$ is true if and only if there is an interpretation in which $\mathscr{A}^S$ is true.

Suppose first that there is an interpretation I in which the *wf*. $\mathscr{A}$ is true.

Then the *wf.* $(\exists x_2) \mathscr{B}(x_1, x_2)$ is true in I also, by Proposition 3.27. This means that every valuation in I satisfies $(\exists x_2) \mathscr{B}(x_1, x_2)$. Let $x \in D_I$, and let v be a valuation in which $v(1) = x$. Then there is a valuation v' in I which is 2-equivalent to v and which satisfies $\mathscr{B}(x_1, x_2)$, by Proposition 3.29. Let $\bar{h}_1^1(x) = v'(2)$. This can be done for each $x \in D_I$, thus defining the values of a function $\bar{h}_1^1$. Extend I to an interpretation I^S of the language which includes the Skolem function h_1^1 by including $\bar{h}_1^1$ as the interpretation of h_1^1. Then every valuation in I^S satisfies $\mathscr{B}(x_1, h_1^1(x_1))$, and so (using Proposition 3.27 again) $(\forall x_1) \mathscr{B}(x_1, h_1^1(x_1))$ is true in I^S.

Now suppose, conversely, that there is an interpretation I^E of the extended language (including h_1^1) in which $\mathscr{A}^S$ is true. This means that every valuation in I^E satisfies $\mathscr{B}(x_1, h_1^1(x_1))$. Let I be an interpretation which is the same as I^E but without the interpretation of h_1^1, and let v be a valuation in I. Construct a valuation v' thus: $v'(2) = \bar{h}_1^1(v(1))$, and $v'(j) = v(j)$ for $j \neq 2$. Then v' is a valuation in I which satisfies $\mathscr{B}(x_1, x_2)$, and which is 2-equivalent to v. By Proposition 3.29, then, v satisfies $(\exists x_2) \mathscr{B}(x_1, x_2)$. Hence all valuations in I satisfy $(\exists x_2) \mathscr{B}(x_1, x_2)$, and so $(\forall x_1)(\exists x_2) \mathscr{B}(x_1, x_2)$ is true in I, as required.

Exercise

24   Skolemise the following *wf*s.

(a) $(\forall x_1)(\exists x_2)(\forall x_3) A_1^3(x_1, x_2, x_3)$.

(b) $(\exists x_1)((\forall x_2) A_1^2(x_1, x_2) \rightarrow (\exists x_3) A_2^2(x_1, x_3))$.

(c) $(\forall x_1)(\sim A_1^1(x_1) \rightarrow (\exists x_2)(\exists x_3)(\sim A_1^2(x_2, x_3)))$.

(d) $(\forall x_1)(\exists x_2)(\forall x_3)(\exists x_4)((\sim A_1^2(x_1, x_2) \vee A_2^1(x_1)) \rightarrow A_2^2(x_3, x_4))$.

4
Formal predicate calculus

4.1 The formal system $K_{\mathscr{L}}$

In Chapter 3 we described the formal languages which we shall use, and we saw how different kinds of statements can be translated into *wfs*. of appropriate first order languages. As we did in Chapter 3, we shall in this chapter consider a fixed but unspecified first order language $\mathscr{L}$, in order that our results may be general and apply to all first order languages. The symbols of $\mathscr{L}$ may be interpreted in many different ways, but we shall now concern ourselves with the purely formal aspects of the language, and consider the logical relationships of *wfs*. rather than properties which depend on particular interpretations. The pattern is similar to that of Chapter 2. We define a formal deductive system, and then we demonstrate that it has the properties that it ought to have, namely that it is consistent and that its class of theorems is just the class of logically valid *wfs*.

Fix a first order language $\mathscr{L}$. Define a formal deductive system $K_{\mathscr{L}}$ by the following axioms and rules of deduction.

Axioms

Let $\mathscr{A}$, $\mathscr{B}$, $\mathscr{C}$ be any *wfs*. of $\mathscr{L}$. The following are axioms of $K_{\mathscr{L}}$.

(K1) $(\mathscr{A} \rightarrow (\mathscr{B} \rightarrow \mathscr{A}))$.

(K2) $(\mathscr{A} \rightarrow (\mathscr{B} \rightarrow \mathscr{C})) \rightarrow ((\mathscr{A} \rightarrow \mathscr{B}) \rightarrow (\mathscr{A} \rightarrow \mathscr{C}))$.

(K3) $(\sim\!\mathscr{A} \rightarrow \sim\!\mathscr{B}) \rightarrow (\mathscr{B} \rightarrow \mathscr{A})$.

(K4) $((\forall x_i)\mathscr{A} \rightarrow \mathscr{A})$, if x_i does not occur free in $\mathscr{A}$.

(K5) $((\forall x_i)\mathscr{A}(x_i) \rightarrow \mathscr{A}(t))$, if $\mathscr{A}(x_i)$ is a *wf*. of $\mathscr{L}$ and t is a term in $\mathscr{L}$ which is free for x_i in $\mathscr{A}(x_i)$.

(K6) $(\forall x_i)(\mathscr{A} \rightarrow \mathscr{B}) \rightarrow (\mathscr{A} \rightarrow (\forall x_i)\mathscr{B})$, if $\mathscr{A}$ contains no free occurrence of the variable x_i.

Notice that these are axiom schemes, each with infinitely many instances.

Rules

(1) *Modus ponens*, i.e. from $\mathscr{A}$ and $(\mathscr{A} \rightarrow \mathscr{B})$ deduce $\mathscr{B}$, where $\mathscr{A}$ and $\mathscr{B}$ are any *wfs*. of $\mathscr{L}$.

(2) Generalisation, i.e. from $\mathscr{A}$ deduce $(\forall x_i)\mathscr{A}$, where $\mathscr{A}$ is any *wf.* of $\mathscr{L}$ and x_i is any variable.

Remark 4.1

(*a*) The axiom schemes and rules of deduction for $K_{\mathscr{L}}$ include the axiom schemes and rule of $\mathscr{L}$. The additional axioms and rule are necessary for proofs which involve properties of quantifiers.

(*b*) Axiom $(K5)$ is stated in its most general form. In applications we shall often come across the instance $((\forall x_i)\mathscr{A} \to \mathscr{A})$, where x_i may or may not be a free variable of $\mathscr{A}$. If x_i appears free in $\mathscr{A}$ we may write $\mathscr{A}$ as $\mathscr{A}(x_i)$, and $(K5)$ gives $((\forall x_i)\mathscr{A}(x_i) \to \mathscr{A}(x_i))$, since x_i is free for x_i in $\mathscr{A}(x_i)$ (Remark 3.13). If x_i does not appear free in $\mathscr{A}$, then $(K4)$ gives us $((\forall x_i)\mathscr{A} \to \mathscr{A})$.

Definition 4.2

(See Definition 2.2). A *proof* in $K_{\mathscr{L}}$ is a sequence of *wfs.* $\mathscr{A}_1, \ldots, \mathscr{A}_n$ of $\mathscr{L}$ such that for each i $(1 \le i \le n)$, either $\mathscr{A}_i$ is an axiom of $K_{\mathscr{L}}$ or $\mathscr{A}_i$ follows from previous members of the sequence by *MP* or Generalisation.

If Γ is a set of *wfs.* of $\mathscr{L}$, a *deduction from* Γ in $K_{\mathscr{L}}$ is a similar sequence, in which members of Γ may be included. (See Definition 2.5.)

A *wf.* $\mathscr{A}$ is a *theorem* of $K_{\mathscr{L}}$ if it is the last member of some sequence which constitutes a proof in $K_{\mathscr{L}}$.

A *wf.* $\mathscr{A}$ is a *consequence in* $K_{\mathscr{L}}$ of the set Γ of *wfs.* if $\mathscr{A}$ is the last member of a sequence which constitutes a deduction from Γ in $K_{\mathscr{L}}$.

We shall write $\underset{K_{\mathscr{L}}}{\vdash} \mathscr{A}$ to denote '$\mathscr{A}$ is a theorem of $K_{\mathscr{L}}$', and $\Gamma \underset{K_{\mathscr{L}}}{\vdash} \mathscr{A}$ to denote '$\mathscr{A}$ is a consequence in $K_{\mathscr{L}}$ of Γ', where Γ is a set of *wfs.* of $K_{\mathscr{L}}$.

For the sake of convenience we shall abbreviate $K_{\mathscr{L}}$ to K unless there is reason to emphasise the particular language being used.

Proposition 4.3

If $\mathscr{A}$ is a *wf.* of $\mathscr{L}$ and $\mathscr{A}$ is a tautology, then $\mathscr{A}$ is a theorem of K. (We shall find, in contrast to the situation of Chapter 2, that the converse of this is false.)

Proof. A *wf.* $\mathscr{A}$ of $\mathscr{L}$ is a tautology if there is a *wf.* $\mathscr{A}_0$ of L from which $\mathscr{A}$ is obtained by substituting *wfs.* of $\mathscr{L}$ for the statement variables, and which is a tautology. Let $\mathscr{A}$ be a *wf.* of $\mathscr{L}$ which is a tautology, and let $\mathscr{A}_0$ be the corresponding *wf.* of L. Then $\mathscr{A}_0$ is a tautology, and so $\underset{L}{\vdash} \mathscr{A}_0$. The proof of $\mathscr{A}_0$ in L can be transformed into a proof of $\mathscr{A}$ in K, simply by replacing statement variables by appropriate *wfs.* of $\mathscr{L}$ throughout. What we obtain is a proof in K because axiom schemes $(L1)$, $(L2)$, $(L3)$

and the rule *MP* are common to the systems L and K. So we have $\underset{K}{\vdash} \mathscr{A}$, as required.

▷ There is a proposition analogous to Proposition 2.14 for the system L. It says that every theorem of K is logically valid. The proof is along similar lines to the proof of 2.14, and we must start by verifying that the axioms of K are all logically valid. All instances of $(K1)$, $(K2)$ and $(K3)$ have been demonstrated to be logically valid (Proposition 3.29), since they are tautologies.

Proposition 4.4

All instances of axiom schemes $(K4)$, $(K5)$ and $(K6)$ are logically valid.

Proof. For $(K4)$, let v be a valuation in an interpretation I of $\mathscr{L}$, and let v satisfy $(\forall x_i)\mathscr{A}$. Then every v' which is i-equivalent to v satisfies $\mathscr{A}$. In particular, v satisfies $\mathscr{A}$. Hence every valuation in I satisfies $((\forall x_i)\mathscr{A} \to \mathscr{A})$, and so $I \vDash ((\forall x_i)\mathscr{A} \to \mathscr{A})$ for every interpretation I, i.e. $((\forall x_i)\mathscr{A} \to \mathscr{A})$ is logically valid.

For $(K5)$, let t be free for x_i in the *wf.* $\mathscr{A}(x_i)$, and let v be a valuation in some interpretation I. If v does not satisfy $(\forall x_i)\mathscr{A}(x_i)$ then v does satisfy $((\forall x_i)\mathscr{A}(x_i) \to \mathscr{A}(t))$ (Definition 3.20). So suppose that v satisfies $(\forall x_i)\mathscr{A}(x_i)$. We show that v satisfies $\mathscr{A}(t)$ also. Every w which is i-equivalent to v satisfies $\mathscr{A}(x_i)$. In particular, v' satisfies $\mathscr{A}(x_i)$, where $v'(x_i) = v(t)$ and $v'(x_k) = v(x_k)$ for $k \neq i$. Therefore, by Proposition 3.23, v satisfies $\mathscr{A}(t)$. It follows now that every valuation v in I satisfies $((\forall x_i)\mathscr{A}(x_i) \to \mathscr{A}(t))$, and so $I \vDash ((\forall x_i)\mathscr{A}(x_i) \to \mathscr{A}(t))$, for every I, i.e. $((\forall x_i)\mathscr{A}(x_i) \to \mathscr{A}(t))$ is logically valid, as required.

For $(K6)$, let $\mathscr{A}$ and $\mathscr{B}$ be *wfs.* of $\mathscr{L}$ and suppose that x_i does not occur free in $\mathscr{A}$. Let v be a valuation in some interpretation I. The verification follows the previous pattern. Suppose that v satisfies $(\forall x_i)(\mathscr{A} \to \mathscr{B})$. Then every w which is i-equivalent to v satisfies $(\mathscr{A} \to \mathscr{B})$. Therefore every such w either does not satisfy $\mathscr{A}$ or does satisfy $\mathscr{B}$. Now if one such w does not satisfy $\mathscr{A}$ then every such w does not satisfy $\mathscr{A}$, since x_i does not occur free in $\mathscr{A}$ (by Proposition 3.33). v is such a w, so we have:

> *either* v does not satisfy $\mathscr{A}$,
>
> *or* every w which is i-equivalent to v satisfies $\mathscr{B}$.

Hence
> *either* v does not satisfy $\mathscr{A}$ or v satisfies $(\forall x_i)\mathscr{B}$.

That is, v satisfies $(\mathscr{A} \to (\forall x_i)\mathscr{B})$. Thus every valuation in I satisfies $(\forall x_i)(\mathscr{A} \to \mathscr{B}) \to (\mathscr{A} \to (\forall x_i)\mathscr{B})$. As previously, then, $(\forall x_i)(\mathscr{A} \to \mathscr{B}) \to (\mathscr{A} \to (\forall x_i)\mathscr{B})$ is logically valid.

Proposition 4.5 (The Soundness Theorem for K)

For any *wf.* $\mathscr{A}$ of $\mathscr{L}$, if $\vdash_K \mathscr{A}$ then $\mathscr{A}$ is logically valid.

Proof. By induction on the number of steps in a proof of $\mathscr{A}$.

Base step: If $\mathscr{A}$ has a one-step proof then $\mathscr{A}$ is an axiom of K. We have shown above that every axiom of K is logically valid.

Induction step: Suppose that $\mathscr{A}$ has a proof with n steps ($n > 1$), and that all theorems of K with proofs in fewer than n steps are logically valid. $\mathscr{A}$ appears in a proof, so either $\mathscr{A}$ is an axiom or $\mathscr{A}$ follows from previous *wfs.* in the proof, using MP or Generalisation. If $\mathscr{A}$ is an axiom then it is logically valid, as above. If $\mathscr{A}$ follows from $\mathscr{B}$ and $(\mathscr{B} \rightarrow \mathscr{A})$, previous *wfs.* in the proof, by MP, then $\mathscr{B}$ and $(\mathscr{B} \rightarrow \mathscr{A})$ have shorter proofs and so by the induction hypothesis are logically valid. By Remark 3.36(a), then, $\mathscr{A}$ is logically valid. Also, if $\mathscr{A}$ follows by Generalisation from a previous *wf.* $\mathscr{C}$, say, then $\mathscr{C}$ is logically valid, and $\mathscr{A}$, being $(\forall x_i)\mathscr{C}$, is also logically valid, by Remark 3.36(b). Hence, in every case, $\mathscr{A}$ is logically valid.

This completes our proof by induction.

Corollary 4.6

K is consistent (i.e. for no *wf.* $\mathscr{A}$ are both $\mathscr{A}$ and $\sim\!\mathscr{A}$ theorems of K).

Proof. Suppose that $\vdash_K \mathscr{A}$ and $\vdash_K (\sim\!\mathscr{A})$, for some *wf.* $\mathscr{A}$ of $\mathscr{L}$. Then $\mathscr{A}$ and $(\sim\!\mathscr{A})$ are both logically valid, by Proposition 4.5. Hence in any interpretation both $\mathscr{A}$ and $(\sim\!\mathscr{A})$ are true.

This contradicts Remark 3.25(c), so K must be consistent.

▷ Finding proofs in K for theorems of K is difficult, as it was for L, and we again look for methods to help demonstrate that particular *wfs.* are theorems. We have a proposition for K which corresponds to the Deduction Theorem (Proposition 2.8), but it is slightly more complicated. Let us see, by means of an example, why this has to be.

Example 4.7

We know that for any *wf.* $\mathscr{A}$ of K, $\mathscr{A} \vdash_K (\forall x_1)\mathscr{A}$ (this is immediate, by the rule Generalisation). However, it is not necessarily the case that $\vdash_K (\mathscr{A} \rightarrow (\forall x_1)\mathscr{A})$.

To see this, let $\mathscr{A}$ be $A_1^1(x_1)$. Let I be an interpretation whose domain is the set $\mathbb{Z}$ of integers, and let $\bar{A}_1^1$ be the predicate '$=0$'. $A_1^1(x_1)$ is interpreted intuitively, then, as '$x = 0$'. Any valuation in I in which $v(x_1) = 0$ will satisfy $A_1^1(x_1)$. However any valuation 1-equivalent to such a v (and different from it) will not satisfy $A_1^1(x_1)$. So no valuation in

I satisfies $(\forall x_1)A_1^1(x_1)$. There is therefore a valuation which satisfies $A_1^1(x_1)$ but does not satisfy $(\forall x_1)A_1^1(x_1)$. This valuation does not satisfy $(A_1^1(x_1) \to (\forall x_1)A_1^1(x_1))$, and so this *wf.* is not true in I. It is therefore not logically valid, and so, by Proposition 4.5, it cannot be a theorem of K.

Proposition 4.8 (The Deduction Theorem for K)

Let $\mathscr{A}$ and $\mathscr{B}$ be *wfs.* of $\mathscr{L}$ and let Γ be a set (possibly empty) of *wfs.* of $\mathscr{L}$. If $\Gamma \cup \{\mathscr{A}\} \vdash_K \mathscr{B}$, and the deduction contains no application of Generalisation involving a variable which occurs free in $\mathscr{A}$, then $\Gamma \vdash_K (\mathscr{A} \to \mathscr{B})$.

Proof. The proof is by induction on n, the number of *wfs.* in the deduction of $\mathscr{B}$ from $\Gamma \cup \{\mathscr{A}\}$.

Base step: $n = 1$. $\mathscr{B}$ is an axiom, or $\mathscr{A}$, or a member of Γ. We deduce that $\Gamma \vdash_K (\mathscr{A} \to \mathscr{B})$ in exactly the same way as we did in the corresponding proof for the Deduction Theorem for L.

Induction step: Let $n > 1$. Suppose that if $\mathscr{F}$ is a *wf.* of $\mathscr{L}$ which can be deduced from $\Gamma \cup \{\mathscr{A}\}$, without using Generalisation applied to a free variable of $\mathscr{A}$, in a deduction containing fewer than n *wfs.*, then $\Gamma \vdash_K (\mathscr{A} \to \mathscr{F})$.

Case 1: $\mathscr{B}$ follows from previous *wfs.* in the deduction, by *MP*. The proof here is again the same as for L.

Case 2: $\mathscr{B}$ is an axiom, or $\mathscr{A}$, or a member of Γ. Again the proof is as for L.

Case 3: $\mathscr{B}$ follows from a previous *wf.* in the deduction, by Generalisation. So $\mathscr{B}$ is $(\forall x_i)\mathscr{C}$, say, and $\mathscr{C}$ appears previously in the deduction. Thus $\Gamma \cup \{\mathscr{A}\} \vdash_K \mathscr{C}$, and the deduction contains fewer than n *wfs.*, so $\Gamma \vdash_K (\mathscr{A} \to \mathscr{C})$, since there is no application of Generalisation involving a free variable of $\mathscr{A}$. Also x_i cannot occur free in $\mathscr{A}$, as it is involved in an application of Generalisation in the deduction of $\mathscr{B}$ from $\Gamma \cup \{\mathscr{A}\}$. So we have a deduction of $(\mathscr{A} \to \mathscr{B})$ from Γ as follows:

(1)

$\vdots$ $\left.\rule{0pt}{40pt}\right\}$ deduction of $(\mathscr{A} \to \mathscr{C})$ from Γ

(k) $(\mathscr{A} \to \mathscr{C})$

$(k+1)$ $(\forall x_i)(\mathscr{A} \to \mathscr{C})$ (k), Generalisation

$(k+2)$ $(\forall x_i)(\mathscr{A} \to \mathscr{C}) \to (\mathscr{A} \to (\forall x_i)\mathscr{C})$ $(K6)$

$(k+3)$ $(\mathscr{A} \to (\forall x_i)\mathscr{C})$ $(k+1), (k+2), MP$.

So $\Gamma \vdash_K (\mathscr{A} \to \mathscr{B})$ as required, and this concludes our inductive proof.

▷ This is the most useful version of the Deduction Theorem for K. It is possible to weaken the additional condition concerning the use of Generalisation (see p. 61 in Mendelson), but we shall not need to. Strengthening this additional condition gives the following Corollary, which is often useful.

Corollary 4.9

If $\Gamma \cup \{\mathscr{A}\} \underset{K}{\vdash} \mathscr{B}$, and $\mathscr{A}$ is a closed $wf.$, then $\Gamma \underset{K}{\vdash} (\mathscr{A} \to \mathscr{B})$.

In spite of the apparently less general form of the Deduction Theorem for K, we can still apply it usefully in showing that certain $wfs.$ are theorems, just as we did for L.

Corollary 4.10

For any $wfs.$ $\mathscr{A}$, $\mathscr{B}$, $\mathscr{C}$ of $\mathscr{L}$,

$$\{(\mathscr{A} \to \mathscr{B}), (\mathscr{B} \to \mathscr{C})\} \underset{K}{\vdash} (\mathscr{A} \to \mathscr{C}).$$

Proof. The proof is identical with the proof of Corollary 2.10.

▷ Note therefore that the rule HS may be legitimately used in K as in L. Just as for L, the Deduction Theorem for K has a converse.

Proposition 4.11

Suppose that $\mathscr{A}$ and $\mathscr{B}$ are $wfs.$ of $\mathscr{L}$, Γ is a set of $wfs.$ of $\mathscr{L}$, and $\Gamma \underset{K}{\vdash} (\mathscr{A} \to \mathscr{B})$. Then $\Gamma \cup \{\mathscr{A}\} \underset{K}{\vdash} \mathscr{B}$.

Proof. Identical with the proof of Proposition 2.9.

Example 4.12

If x_i does not occur free in $\mathscr{A}$, then

$$\underset{K}{\vdash} ((\mathscr{A} \to (\forall x_i)\mathscr{B}) \to (\forall x_i)(\mathscr{A} \to \mathscr{B})).$$

We write out a deduction.

(1)	$(\mathscr{A} \to (\forall x_i)\mathscr{B})$	assumption
(2)	$(\forall x_i)\mathscr{B} \to \mathscr{B}$	$(K4)$ or $(K5)$
(3)	$(\mathscr{A} \to \mathscr{B})$	(1), (2), HS
(4)	$(\forall x_i)(\mathscr{A} \to \mathscr{B})$	(3), Generalisation.

So we have

$$(\mathscr{A} \to (\forall x_i)\mathscr{B}) \underset{K}{\vdash} (\forall x_i)(\mathscr{A} \to \mathscr{B}).$$

Now Generalisation is used in the deduction, but only using the variable x_i, which does not occur free in $(\mathscr{A} \to (\forall x_i)\mathscr{B})$. Hence we can apply the Deduction Theorem, to obtain

$$\underset{K}{\vdash} ((\mathscr{A} \to (\forall x_i)\mathscr{B}) \to (\forall x_i)(\mathscr{A} \to \mathscr{B})),$$

as required.

Example 4.13

For any *wfs.* $\mathscr{A}$, $\mathscr{B}$ of $\mathscr{L}$,

$$\underset{K}{\vdash} ((\forall x_i)(\mathscr{A} \to \mathscr{B}) \to ((\exists x_i)\mathscr{A} \to (\exists x_i)\mathscr{B})).$$

Again we write out a deduction. Step (2) in this deduction is not at first sight the obvious one, but we shall see the reason for it at the end.

(1)	$(\forall x_i)(\mathscr{A} \to \mathscr{B})$	assumption
(2)	$(\forall x_i)(\sim\mathscr{B})$	assumption
(3)	$(\forall x_i)(\mathscr{A} \to \mathscr{B}) \to (\mathscr{A} \to \mathscr{B})$	$(K4)$ or $(K5)$
(4)	$(\mathscr{A} \to \mathscr{B})$	$(1), (3), MP$
(5)	$(\mathscr{A} \to \mathscr{B}) \to (\sim\mathscr{B} \to \sim\mathscr{A})$	tautology
(6)	$(\sim\mathscr{B} \to \sim\mathscr{A})$	$(4), (5), MP$
(7)	$((\forall x_i)(\sim\mathscr{B}) \to (\sim\mathscr{B}))$	$(K4)$ or $(K5)$
(8)	$(\sim\mathscr{B})$	$(2), (7), MP$
(9)	$(\sim\mathscr{A})$	$(6), (8), MP$
(10)	$(\forall x_i)(\sim\mathscr{A})$	$(9),$ Generalisation.

This demonstrates that

$$\{(\forall x_1)(\mathscr{A} \to \mathscr{B}), (\forall x_i)(\sim\mathscr{B})\} \underset{K}{\vdash} (\forall x_i)(\sim\mathscr{A}).$$

By the Deduction Theorem, we get

$$(\forall x_i)(\mathscr{A} \to \mathscr{B}) \underset{K}{\vdash} ((\forall x_i)(\sim\mathscr{B}) \to (\forall x_i)(\sim\mathscr{A})),$$

since x_i does not occur free in $(\forall x_i)(\sim\mathscr{B})$. Now we know that

$$\underset{K}{\vdash} ((\forall x_i)(\sim\mathscr{B}) \to (\forall x_i)(\sim\mathscr{A})) \to (\sim(\forall x_i)(\sim\mathscr{A}) \to \sim(\forall x_i)(\sim\mathscr{B}))$$

and so, using *MP*, we have

$$(\forall x_i)(\mathscr{A} \to \mathscr{B}) \underset{K}{\vdash} (\sim(\forall x_i)(\sim\mathscr{A}) \to \sim(\forall x_i)(\sim\mathscr{B}))$$

i.e.

$$(\forall x_i)(\mathscr{A} \to \mathscr{B}) \underset{K}{\vdash} ((\exists x_i)\mathscr{A} \to (\exists x_i)\mathscr{B}).$$

Again, using the Deduction Theorem,

$$\underset{K}{\vdash} ((\forall x_i)(\mathcal{A} \rightarrow \mathcal{B}) \rightarrow ((\exists x_i)\mathcal{A} \rightarrow (\exists x_i)\mathcal{B})),$$

since x_i is not free in $(\forall x_i)(\mathcal{A} \rightarrow \mathcal{B})$.

Exercises

1   Write out a proof in $K_{\mathcal{L}}$ of the *wf*.

$$(\forall x_1)(A_1^1(x_1) \rightarrow A_1^1(x_1)).$$

2   Prove that the following are theorems of $K_{\mathcal{L}}$.
       (a) $(\exists x_i)(\mathcal{A} \rightarrow \mathcal{B}) \rightarrow ((\forall x_i)\mathcal{A} \rightarrow \mathcal{B})$,
       (b) $((\exists x_i)\mathcal{A} \rightarrow \mathcal{B}) \rightarrow (\forall x_i)(\mathcal{A} \rightarrow \mathcal{B})$, provided that x_i does not occur free in $\mathcal{B}$.
       (c) $(\sim(\forall x_i)\mathcal{A} \rightarrow (\exists x_i)\sim\mathcal{A})$.

3(a)   What is wrong with the following?
       (1) $(\exists x_2)A_1^2(x_1, x_2)$ assumption
       (2) $(\forall x_1)(\exists x_2)A_1^2(x_1, x_2)$ (1), Generalisation
       (3) $(\forall x_1)(\exists x_2)A_1^2(x_1, x_2) \rightarrow (\exists x_2)A_1^2(x_2, x_2)$ $(K5)$
       (4) $(\exists x_2)A_1^2(x_2, x_2)$ (2), (3), *MP*.

Therefore, $(\exists x_2)A_1^2(x_1, x_2) \underset{K}{\vdash} (\exists x_2)A_1^2(x_2, x_2)$, and hence by the Deduction Theorem,

$$\underset{K}{\vdash} (\exists x_2)A_1^2(x_1, x_2) \rightarrow (\exists x_2)A_1^2(x_2, x_2).$$

(b) Show, by finding a suitable interpretation, that the formula $((\exists x_2)A_1^2(x_1, x_2) \rightarrow (\exists x_2)A_1^2(x_2, x_2))$ is not logically valid, and is therefore not a theorem of K.

4.2 Equivalence, substitution

Remark 4.14

It is convenient to introduce the connective $\leftrightarrow$ as a defined symbol into our language. For *wfs*. $\mathcal{A}$ and $\mathcal{B}$ of $\mathcal{L}$, $(\mathcal{A} \leftrightarrow \mathcal{B})$ is to stand for $\sim((\mathcal{A} \rightarrow \mathcal{B}) \rightarrow \sim(\mathcal{B} \rightarrow \mathcal{A}))$. Note again that $(\mathcal{A} \leftrightarrow \mathcal{B})$ is not a *wf*. of $\mathcal{L}$, but we use it as a convenient abbreviation for a particular *wf*.

Proposition 4.15

For any *wfs*. $\mathcal{A}$ and $\mathcal{B}$ of $\mathcal{L}$, $\underset{K}{\vdash} (\mathcal{A} \leftrightarrow \mathcal{B})$ if and only if $\underset{K}{\vdash} (\mathcal{A} \rightarrow \mathcal{B})$ and $\underset{K}{\vdash} (\mathcal{B} \rightarrow \mathcal{A})$.

Proof. First suppose that $\underset{K}{\vdash} (\mathcal{A} \leftrightarrow \mathcal{B})$, i.e. $\underset{K}{\vdash} \sim((\mathcal{A} \rightarrow \mathcal{B}) \rightarrow \sim(\mathcal{B} \rightarrow \mathcal{A}))$. Now the *wfs*. $(\sim((\mathcal{A} \rightarrow \mathcal{B}) \rightarrow \sim(\mathcal{B} \rightarrow \mathcal{A})) \rightarrow (\mathcal{A} \rightarrow \mathcal{B}))$ and $(\sim((\mathcal{A} \rightarrow \mathcal{B}) \rightarrow$

$\sim(\mathcal{B} \to \mathcal{A})) \to (\mathcal{B} \to \mathcal{A}))$ are tautologies (verification left to the reader), and so by Proposition 4.3 they are theorems of K. By MP, then, we have

$$\underset{K}{\vdash} (\mathcal{A} \to \mathcal{B}) \quad \text{and} \quad \underset{K}{\vdash} (\mathcal{B} \to \mathcal{A})$$

as required.

Now suppose that $\underset{K}{\vdash} (\mathcal{A} \to \mathcal{B})$ and $\underset{K}{\vdash} (\mathcal{B} \to \mathcal{A})$. We must show that $\underset{K}{\vdash} \sim((\mathcal{A} \to \mathcal{B}) \to \sim(\mathcal{B} \to \mathcal{A}))$. It is sufficient to show that the *wf.*

$$(\mathcal{A} \to \mathcal{B}) \to ((\mathcal{B} \to \mathcal{A}) \to \sim((\mathcal{A} \to \mathcal{B}) \to \sim(\mathcal{B} \to \mathcal{A})))$$

is a tautology, and this can be done easily by constructing a truth table.

Definition 4.16

If $\mathcal{A}$ and $\mathcal{B}$ are *wfs.* of $\mathcal{L}$ and $\underset{K}{\vdash} (\mathcal{A} \leftrightarrow \mathcal{B})$, we say that $\mathcal{A}$ and $\mathcal{B}$ are *provably equivalent*.

Corollary 4.17

For any *wfs.* $\mathcal{A}$, $\mathcal{B}$, $\mathcal{C}$ of $\mathcal{L}$, if $\mathcal{A}$ and $\mathcal{B}$ are provably equivalent and $\mathcal{B}$ and $\mathcal{C}$ are provably equivalent, then $\mathcal{A}$ and $\mathcal{C}$ are provably equivalent.

Proof. Let $\underset{K}{\vdash} (\mathcal{A} \leftrightarrow \mathcal{B})$ and $\underset{K}{\vdash} (\mathcal{B} \leftrightarrow \mathcal{C})$. Then $\underset{K}{\vdash} (\mathcal{A} \to \mathcal{B})$ and $\underset{K}{\vdash} (\mathcal{B} \to \mathcal{C})$, so by HS we have $\underset{K}{\vdash} (\mathcal{A} \to \mathcal{C})$.

Also $\underset{K}{\vdash} (\mathcal{B} \to \mathcal{A})$ and $\underset{K}{\vdash} (\mathcal{C} \to \mathcal{B})$, so again by HS we have $\underset{K}{\vdash} (\mathcal{C} \to \mathcal{A})$. Hence, by Proposition 4.15, $\underset{K}{\vdash} (\mathcal{A} \leftrightarrow \mathcal{C})$.

$\triangleright$ This last proposition will be useful when we need to show that two *wfs.* are provably equivalent. We merely have to show that both implications are theorems. We shall require to do this in the next part of our description of K, in which we investigate how parts of *wfs.* may be substituted and how variables may be substituted. We did this in Chapter 1 in an informal way. The presence of variables complicates things here, so some of the proofs are lengthy. However, the results will be used subsequently, so this is a necessary part of our exposition.

We start with consideration of how variables may be substituted. The *wf.* $(\forall x_1)A_1^1(x_1)$ will be interpreted (intuitively) as 'for all x, $\bar{A}_1^1(x)$ holds'. Likewise the *wf.* $(\forall x_2)A_1^1(x_2)$ will be interpreted intuitively as 'for all x, $\bar{A}_1^1(x)$ holds'. So it would seem that the actual variable which appears in the *wf.* (in this case) does not affect the interpretation. In the formal system K, therefore, these two *wfs.* ought to be equivalent, in some sense. This is made precise in the next proposition.

As before, let $\mathcal{A}(x_i)$ denote a *wf.* of $\mathcal{L}$ in which x_i occurs free (possibly more, than once). Then for any variable x_j, we denote by $\mathcal{A}(x_j)$ the *wf.* obtained by substituting x_j for each free occurrence of x_i in $\mathcal{A}(x_i)$.

Proposition 4.18

If x_i occurs free in $\mathcal{A}(x_i)$ and x_j is a variable which does not occur, free or bound, in $\mathcal{A}(x_i)$, then

$$\underset{K}{\vdash} ((\forall x_i)\mathcal{A}(x_i) \leftrightarrow (\forall x_j)\mathcal{A}(x_j)).$$

Proof. First observe that under the conditions specified, x_i is free for x_j in $\mathcal{A}(x_j)$ and x_j is free for x_i in $\mathcal{A}(x_i)$. We need two deductions, to demonstrate that both implications are theorems of K.

(1)	$(\forall x_i)\mathcal{A}(x_i)$	assumption
(2)	$((\forall x_i)\mathcal{A}(x_i) \to \mathcal{A}(x_j))$	$(K5)$
(3)	$\mathcal{A}(x_j)$	(1), (2), MP
(4)	$(\forall x_j)\mathcal{A}(x_j)$	(3), Generalisation
∴	$(\forall x_i)\mathcal{A}(x_i) \underset{K}{\vdash} (\forall x_j)\mathcal{A}(x_j).$	

Hence, by the Deduction Theorem,

$$\underset{K}{\vdash} ((\forall x_i)\mathcal{A}(x_i) \to (\forall x_j)\mathcal{A}(x_j)),$$

since x_j is not free in $(\forall x_i)\mathcal{A}(x_i)$. In precisely the same way we prove that

$$\underset{K}{\vdash} ((\forall x_j)\mathcal{A}(x_j) \to (\forall x_i)\mathcal{A}(x_i)).$$

Hence, by Proposition 4.15, we have

$$\underset{K}{\vdash} ((\forall x_i)\mathcal{A}(x_i) \leftrightarrow (\forall x_j)\mathcal{A}(x_j)).$$

▷ This proposition shows that we may replace a particular bound variable to obtain a *wf.* which is provably equivalent to the original *wf.*, provided that we choose the new variable appropriately. The usefulness of the procedure will become apparent later.

Proposition 4.19

Let $\mathcal{A}$ be a *wf.* of $\mathcal{L}$ whose free variables are $y_1, \ldots, y_n$. Then $\underset{K}{\vdash} \mathcal{A}$ if and only if $\underset{K}{\vdash} (\forall y_1) \ldots (\forall y_n)\mathcal{A}$.

Proof. Suppose first that $\underset{K}{\vdash} \mathcal{A}$. We proceed by induction on n, the number of free variables in $\mathcal{A}$.

Base step: $n = 1$ (the case of formulas with no free variables is trivial). $\mathscr{A}$ has one free variable, y_1. If $\vdash_K \mathscr{A}(y_1)$ then $\vdash_K (\forall y_1)\mathscr{A}(y_1)$, by a single application of Generalisation.

Induction step: Let $n > 1$, and suppose that the result is true for every wf. of $\mathscr{L}$ with $n-1$ free variables. Consider the wf. $(\forall y_n)\mathscr{A}$. This has $n-1$ free variables. We have $\vdash_K \mathscr{A}$, so $\vdash_K (\forall y_n)\mathscr{A}$, by Generalisation, so $\vdash_K (\forall y_1)\ldots(\forall y_{n-1})(\forall y_n)\mathscr{A}$, by the induction hypothesis.

Conversely, suppose that $\vdash_K (\forall y_1)\ldots(\forall y_n)\mathscr{A}$. We prove that $\vdash_K \mathscr{A}$ similarly by induction on n, using applications of axiom $(K5)$.

Definition 4.20

If $\mathscr{A}$ is a wf. of $\mathscr{L}$ containing free occurrences of the variables $y_1,\ldots,y_n$ only, then wf. $(\forall y_1)\ldots(\forall y_n)\mathscr{A}$ is the *universal closure* of $\mathscr{A}$. The universal closure of $\mathscr{A}$ is usually denoted by $\mathscr{A}'$.

Remark 4.21

The above proposition says that for any wf. $\mathscr{A}$ of $\mathscr{L}$, $\vdash_K \mathscr{A}$ if and only if $\vdash_K \mathscr{A}'$. We should take care to remember, however that $\mathscr{A}$ and $\mathscr{A}'$ are not in general provably equivalent. It is not hard to show that $\vdash_K (\mathscr{A}' \to \mathscr{A})$ always holds, but we have seen in Example 4.7 that $(\mathscr{A} \to \mathscr{A}')$ is not necessarily a theorem of K.

Proposition 4.22

Let $\mathscr{A}$ and $\mathscr{B}$ be wfs. of $\mathscr{L}$, and suppose that $\mathscr{B}_0$ arises from the wf. $\mathscr{A}_0$ by substituting $\mathscr{B}$ for one or more occurrences of $\mathscr{A}$ in $\mathscr{A}_0$. Then

$$\vdash_K ((\mathscr{A} \leftrightarrow \mathscr{B})' \to (\mathscr{A}_0 \leftrightarrow \mathscr{B}_0)).$$

Proof. The proof is by induction on the length of (i.e. the number of connectives and quantifiers in) $\mathscr{A}_0$.

Base step: We are assuming, necessarily, that $\mathscr{A}_0$ contains $\mathscr{A}$ as a subformula. $\mathscr{A}_0$ has fewest connectives and quantifiers, therefore, when $\mathscr{A}_0$ is $\mathscr{A}$ itself. In this case, $\mathscr{B}_0$ is just $\mathscr{B}$. Then $\vdash_K ((\mathscr{A} \leftrightarrow \mathscr{B})' \to (\mathscr{A} \leftrightarrow \mathscr{B}))$ is an instance of a general result mentioned in Remark 4.21 above.

Induction step: Suppose that $\mathscr{A}_0$ contains $\mathscr{A}$ as a strict subformula, and that the result is true for all wfs. which are shorter than $\mathscr{A}_0$ and which contain $\mathscr{A}$ as a subformula. As in previous proofs, there are three cases to consider.

Case 1: $\mathscr{A}_0$ is $\sim\mathscr{C}_0$. Then $\mathscr{B}_0$ is $\sim\mathscr{D}_0$, say, where $\mathscr{D}_0$ is the result of substituting $\mathscr{B}$ for $\mathscr{A}$ in $\mathscr{C}_0$. Now $\mathscr{C}_0$ has fewer connectives and quantifiers than $\mathscr{A}_0$, so

$$\vdash_K ((\mathscr{A} \leftrightarrow \mathscr{B})' \to (\mathscr{C}_0 \leftrightarrow \mathscr{D}_0)).$$

Since $((\mathscr{C}_0 \leftrightarrow \mathscr{D}_0) \to (\sim\mathscr{C}_0 \leftrightarrow \sim\mathscr{D}_0))$ is a tautology, it is a theorem of K, and we have, by HS (Corollary 4.10),

$$\vdash_K ((\mathscr{A} \leftrightarrow \mathscr{B})' \to (\sim\mathscr{C}_0 \leftrightarrow \sim\mathscr{D}_0)),$$

i.e.

$$\vdash_K ((\mathscr{A} \leftrightarrow \mathscr{B})') \to (\mathscr{A}_0 \leftrightarrow \mathscr{B}_0)).$$

Case 2: $\mathscr{A}_0$ is $(\mathscr{C}_0 \to \mathscr{D}_0)$. Then $\mathscr{B}_0$ is $(\mathscr{E}_0 \to \mathscr{F}_0)$, say, where $\mathscr{E}_0$ and $\mathscr{F}_0$ are the results of substituting $\mathscr{B}$ for $\mathscr{A}$ in $\mathscr{C}_0$ and $\mathscr{D}_0$ respectively. Now $\mathscr{C}_0$ and $\mathscr{D}_0$ each has fewer connectives and quantifiers than $\mathscr{A}_0$, so

$$\vdash_K ((\mathscr{A} \leftrightarrow \mathscr{B})' \to (\mathscr{C}_0 \leftrightarrow \mathscr{E}_0)),$$

and

$$\vdash_K ((\mathscr{A} \leftrightarrow \mathscr{B})' \to (\mathscr{D}_0 \leftrightarrow \mathscr{F}_0)).$$

It is left as an exercise to verify that it follows from these that

$$\vdash_K ((\mathscr{A} \leftrightarrow \mathscr{B})' \to ((\mathscr{C}_0 \to \mathscr{D}_0) \leftrightarrow (\mathscr{E}_0 \to \mathscr{F}_0))),$$

i.e.

$$\vdash_K ((\mathscr{A} \leftrightarrow \mathscr{B})' \to (\mathscr{A}_0 \leftrightarrow \mathscr{B}_0)).$$

Case 3: $\mathscr{A}_0$ is $(\forall x_i)\mathscr{C}_0$. Then $\mathscr{B}_0$ is $(\forall x_i)\mathscr{D}_0$, say, where $\mathscr{D}_0$ is the result of substituting $\mathscr{B}$ for $\mathscr{A}$ in $\mathscr{C}_0$. Now $\mathscr{C}_0$ has fewer connectives and quantifiers than $\mathscr{A}_0$, so

$$\vdash_K ((\mathscr{A} \leftrightarrow \mathscr{B})' \to (\mathscr{C}_0 \leftrightarrow \mathscr{D}_0)).$$

By Generalisation, then, we get

$$\vdash_K (\forall x_i)((\mathscr{A} \leftrightarrow \mathscr{B})' \to (\mathscr{C}_0 \leftrightarrow \mathscr{D}_0)).$$

Now x_i does not occur free in $(\mathscr{A} \leftrightarrow \mathscr{B})'$, so as an instance of axiom $(K6)$ we have

$$\vdash_K ((\forall x_i)((\mathscr{A} \leftrightarrow \mathscr{B})' \to (\mathscr{C}_0 \leftrightarrow \mathscr{D}_0))$$
$$\to ((\mathscr{A} \leftrightarrow \mathscr{B})' \to (\forall x_i)(\mathscr{C}_0 \leftrightarrow \mathscr{D}_0))).$$

Therefore, by MP,

$$\vdash_K ((\mathscr{A} \leftrightarrow \mathscr{B})' \to (\forall x_i)(\mathscr{C}_0 \leftrightarrow \mathscr{D}_0)),$$

and we obtain our result

$$\underset{K}{\vdash} ((\mathscr{A} \leftrightarrow \mathscr{B})' \to ((\forall x_i)\mathscr{C}_0 \leftrightarrow (\forall x_i)\mathscr{D}_0)),$$

i.e.

$$\underset{K}{\vdash} ((\mathscr{A} \leftrightarrow \mathscr{B})' \to (\mathscr{A}_0 \leftrightarrow \mathscr{B}_0)),$$

by applying the following lemma, whose proof is left as an exercise.
   Lemma: If $\mathscr{A}$ and $\mathscr{B}$ are *wfs.* of $\mathscr{L}$, then

$$\underset{K}{\vdash} (\forall x_i)(\mathscr{A} \leftrightarrow \mathscr{B}) \to ((\forall x_i)\mathscr{A} \leftrightarrow (\forall x_i)\mathscr{B}).$$

This completes our inductive proof.

Corollary 4.23

Let $\mathscr{A}$, $\mathscr{B}$, $\mathscr{A}_0$, $\mathscr{B}_0$ be as in Proposition 4.22 above. If $\underset{K}{\vdash} (\mathscr{A} \leftrightarrow \mathscr{B})$ then
$\underset{K}{\vdash} (\mathscr{A}_0 \leftrightarrow \mathscr{B}_0)$.

Proof. Suppose that $\underset{K}{\vdash} (\mathscr{A} \leftrightarrow \mathscr{B})$. Then $\underset{K}{\vdash} (\mathscr{A} \leftrightarrow \mathscr{B})'$, by Proposition 4.19.
By Proposition 4.22, $\underset{K}{\vdash} ((\mathscr{A} \leftrightarrow \mathscr{B})' \to (\mathscr{A}_0 \leftrightarrow \mathscr{B}_0))$. By *MP*, therefore, we
have $\underset{K}{\vdash} (\mathscr{A}_0 \leftrightarrow \mathscr{B}_0)$.

Corollary 4.24

If x_j does not appear (free or bound) in the *wf.* $\mathscr{A}(x_i)$, and the *wf.* $\mathscr{B}_0$
arises from $\mathscr{A}_0$ by replacing one or more occurrences of $(\forall x_i)\mathscr{A}(x_i)$ by
occurrences of $(\forall x_j)\mathscr{A}(x_j)$, then $\underset{K}{\vdash} (\mathscr{A}_0 \leftrightarrow \mathscr{B}_0)$.

Proof. Just apply Proposition 4.18 and Corollary 4.23.

Exercises

4 Prove that

$$\underset{K}{\vdash} ((\forall x_i)(\mathscr{A} \to \mathscr{B}) \to ((\forall x_i)\mathscr{A} \to (\forall x_i)\mathscr{B})),$$

   for any *wfs.* $\mathscr{A}$ and $\mathscr{B}$.
5 Show that the formulas $\sim(\exists x_i)\mathscr{A}$ and $(\forall x_i)(\sim\mathscr{A})$ are provably equivalent in
   K, for any *wf.* $\mathscr{A}$ of $\mathscr{L}$.
6 Prove carefully that

   (a) $(\forall x_1)(\forall x_2)A_1^2(x_1, x_2) \underset{K}{\vdash} (\forall x_2)(\forall x_3)A_1^2(x_2, x_3)$.

   (b) $(\forall x_1)(\forall x_2)A_1^2(x_1, x_2) \underset{K}{\vdash} (\forall x_1)A_1^2(x_1, x_1)$.

7 Let $\mathscr{A}(x_i)$ be a *wf.* of $\mathscr{L}$ in which x_i occurs free and let x_j be a variable which does not occur, free or bound, in $\mathscr{A}(x_i)$. Prove that

$$\underset{K}{\vdash} ((\exists x_i)\mathscr{A}(x_i) \leftrightarrow (\exists x_j)\mathscr{A}(x_j)).$$

4.3 Prenex form

In Chapter 1 the idea of normal forms was introduced and disjunctive and conjunctive normal forms were discussed. One use that normal forms have is that they bring out relationships in logical structure which may not be obvious in the original formulas. We are now in a position to describe a normal form for *wfs.* of $\mathscr{L}$, and where previously in a normal form we allowed only certain connectives to be used in a standard way, here we are concerned with the arrangement of the quantifiers.

Proposition 4.25

Let $\mathscr{A}$ and $\mathscr{B}$ be *wfs.* of $\mathscr{L}$.
  (i) If x_i does not occur free in $\mathscr{A}$, then

$$\underset{K}{\vdash} ((\forall x_i)(\mathscr{A} \to \mathscr{B}) \leftrightarrow (\mathscr{A} \to (\forall x_i)\mathscr{B})),$$

and

$$\underset{K}{\vdash} ((\exists x_i)(\mathscr{A} \to \mathscr{B}) \leftrightarrow (\mathscr{A} \to (\exists x_i)\mathscr{B})).$$

  (ii) If x_i does not occur free in $\mathscr{B}$, then

$$\underset{K}{\vdash} ((\forall x_i)(\mathscr{A} \to \mathscr{B}) \leftrightarrow ((\exists x_i)\mathscr{A} \to \mathscr{B})),$$

and

$$\underset{K}{\vdash} ((\exists x_i)(\mathscr{A} \to \mathscr{B}) \leftrightarrow ((\forall x_i)\mathscr{A} \to \mathscr{B})).$$

Proof. Eight proofs in K are required. One is trivial, as the *wf.* concerned is an instance of axiom $(K6)$, one has been given already in Example 4.12, and another follows easily from Example 4.13. The others require similar proofs, making repeated use of the Deduction Theorem. They are left as exercises.

Example 4.26

Show that the *wf.*

$$(\forall x_1)A_1^1(x_1) \to (\forall x_2)(\exists x_3)A_1^2(x_2, x_3)$$

is provably equivalent to the *wf.*

$$(\exists x_1)(\forall x_2)(\exists x_3)(A_1^1(x_1) \to A_1^2(x_2, x_3)).$$

We write down a sequence of *wfs.*, each one provably equivalent to the next, using one part of Proposition 4.25 at each stage.

$$(\forall x_1)A_1^1(x_1) \to (\forall x_2)(\exists x_3)A_1^2(x_2, x_3),$$

$$(\exists x_1)(A_1^1(x_1) \to (\forall x_2)(\exists x_3)A_1^2(x_2, x_3)),$$

$$(\exists x_1)(\forall x_2)(A_1^1(x_1) \to (\exists x_3)A_1^2(x_2, x_3)),$$

$$(\exists x_1)(\forall x_2)(\exists x_3)(A_1^1(x_1) \to A_1^2(x_2, x_3)).$$

▷ *Wfs.* of $\mathscr{L}$ can be very complicated, and quantifiers within them can be difficult to relate intuitively, especially if they are separated. We can use the above results about substitution and equivalence to show that every *wf.* is provably equivalent to one in which all the quantifiers appear at the beginning.

Definition 4.27

A *wf.* $\mathscr{A}$ of $\mathscr{L}$ is said to be in *prenex form* if it is

$$(Q_1 x_{i_1})(Q_2 x_{i_2}) \dots (Q_k x_{i_k})\mathscr{D},$$

where $\mathscr{D}$ is a *wf.* of $\mathscr{L}$ with no quantifiers, and each Q_j is either $\forall$ or $\exists$. (A *wf.* with no quantifiers is regarded as a trivial case of a *wf.* in prenex form.)

Proposition 4.28

For any *wf.* $\mathscr{A}$ of $\mathscr{L}$, there is a *wf.* $\mathscr{B}$ which is in prenex form and is provably equivalent to $\mathscr{A}$.

Proof. By Proposition 4.18, we can change all the bound variables of $\mathscr{A}$ so as to make them all different from all the free variables of $\mathscr{A}$ (but leave the free variables unaltered), to obtain a *wf.* $\mathscr{A}_1$, say, such that $\vdash_K (\mathscr{A}_1 \leftrightarrow \mathscr{A})$. Now we proceed by induction on the length of $\mathscr{A}_1$, i.e. the number of connectives and quantifiers in $\mathscr{A}_1$.

Base step: $\mathscr{A}_1$ is an atomic formula. Here there is nothing to prove, for $\mathscr{A}_1$ is already trivially in prenex form.

Induction step: Let $\mathscr{A}_1$ not be an atomic formula, and suppose that every *wf.* which is shorter than $\mathscr{A}_1$ is provably equivalent to a *wf.* in prenex form. There are three cases.

Case 1: $\mathscr{A}_1$ is $\sim\mathscr{C}$. Then $\mathscr{C}$ is shorter than $\mathscr{A}_1$, so there is a *wf.* $\mathscr{C}_1$ in prenex form such that $\vdash_K (\mathscr{C}_1 \leftrightarrow \mathscr{C})$. It follows that $\vdash_K (\mathscr{A}_1 \leftrightarrow (\sim\mathscr{C}_1))$

i.e. $\quad \vdash_K (\mathscr{A}_1 \leftrightarrow \sim(Q_1 x_{i_1}) \dots (Q_k x_{i_k})\mathscr{D})$, say.

So $\vdash_K (\mathscr{A}_1 \leftrightarrow (Q_1^* x_{i_1}) \dots (Q_k^* x_{i_k})(\sim\mathscr{D}))$, where if Q_j is $\forall$ then Q_1^* is $\exists$ and if Q_j is $\exists$ then Q_j^* is $\forall$, for $1 \leq j \leq k$. Let $\mathcal{B}$ be the $wf.$ $(Q_1^* x_{i_1}) \ldots (Q_k^* x_{i_k})(\sim\mathcal{D})$. Then $\underset{K}{\vdash} (\mathcal{A} \leftrightarrow \mathcal{B})$, and $\mathcal{B}$ is in prenex form.

Case 2: $\mathcal{A}_1$ is $(\mathcal{C} \rightarrow \mathcal{D})$. Then $\mathcal{C}$ and $\mathcal{D}$ are shorter than $\mathcal{A}_1$ and so there are $wfs.$ $\mathcal{C}_1$ and $\mathcal{D}_1$ in prenex form such that

$$\underset{K}{\vdash} (\mathcal{C}_1 \leftrightarrow \mathcal{C}) \quad \text{and} \quad \underset{K}{\vdash} (\mathcal{D}_1 \leftrightarrow \mathcal{D}).$$

By Corollary 4.23, then

$$\underset{K}{\vdash} (\mathcal{C} \rightarrow \mathcal{D}) \leftrightarrow (\mathcal{C}_1 \rightarrow \mathcal{D}),$$

and hence, by the same corollary, and Corollary 4.17,

$$\underset{K}{\vdash} (\mathcal{C} \rightarrow \mathcal{D}) \leftrightarrow (\mathcal{C}_1 \rightarrow \mathcal{D}_1),$$

i.e.

$$\underset{K}{\vdash} (\mathcal{A}_1 \leftrightarrow (\mathcal{C}_1 \rightarrow \mathcal{D}_1)).$$

Hence $\underset{K}{\vdash} (\mathcal{A} \leftrightarrow (\mathcal{C}_1 \rightarrow \mathcal{D}_1))$, by Corollary 4.17. Now $(\mathcal{C}_1 \rightarrow \mathcal{D}_1)$ has the form

$$((Q_1 x_{i_1}) \ldots (Q_k x_{i_k})\mathcal{C}_2 \rightarrow (R_1 x_{j_1}) \ldots (R_l x_{j_l})\mathcal{D}_2),$$

where $\mathcal{C}_2$ and $\mathcal{D}_2$ contain no quantifiers, and the Qs and Rs are either $\forall$ or $\exists$. We now apply Proposition 4.25 repeatedly to bring all the quantifiers to the beginning, changed if necessary. We can do this since the $x_{i_1}, \ldots, x_{i_k}, x_{j_1}, \ldots, x_{j_k}$ are all different and different from any of the free variables occurring in $\mathcal{C}_2$ and $\mathcal{D}_2$. We obtain

$$\underset{K}{\vdash} ((\mathcal{C}_1 \rightarrow \mathcal{D}_1) \leftrightarrow (Q_1^* x_{i_1}) \ldots (Q_k^* x_{i_k})$$
$$(R_1 x_{j_1}) \ldots (R_l x_{j_l})(\mathcal{C}_2 \rightarrow \mathcal{D}_2)).$$

This last part is a $wf.$ in prenex form, so is the required $\mathcal{B}$.

Case 3: $\mathcal{A}_1$ is $(\forall x_i)\mathcal{C}$. Then $\mathcal{C}$ is shorter than $\mathcal{A}_1$, and there is a $wf.$ in prenex form provably equivalent to $\mathcal{C}$. Say

$$\underset{K}{\vdash} (\mathcal{C} \leftrightarrow (Q_1 x_{i_1}) \ldots (Q_k x_{i_k})\mathcal{D}).$$

Then, by Generalisation,

$$\underset{K}{\vdash} (\forall x_i)(\mathcal{C} \leftrightarrow (Q_1 x_{i_1}) \ldots (Q_k x_{i_k})\mathcal{D}),$$

and $\underset{K}{\vdash} ((\forall x_i)\mathcal{C} \leftrightarrow (\forall x_i)(Q_1 x_{i_1}) \ldots (Q_k x_{i_k})\mathcal{D})$ follows as in the proof of Proposition 4.22. The $wf.$ $(\forall x_i)(Q_1 x_{i_1}) \ldots (Q_k x_{i_k})\mathcal{D}$ is therefore the required $\mathcal{B}$.

This completes our inductive proof.

Example 4.29

(a) Find a *wf.* in prenex form which is provably equivalent to the *wf.*

$$A_1^1(x_1) \to (\forall x_2)A_1^2(x_1, x_2).$$

This example corresponds with Case 2 in the inductive proof above. First observe that x_2 is the only bound variable and that x_2 does not occur free anywhere, so we need not change any variables. We can apply Proposition 4.25(i) directly to see that

$$(\forall x_2)(A_1^1(x_1) \to A_1^2(x_1, x_2))$$

is provably equivalent to the given *wf.*, and it is in prenex form.

(b) Find a *wf.* in prenex form which is provably equivalent to the *wf.*

$$(((\forall x_1)A_1^2(x_1, x_2) \to {\sim}(\exists x_2)(A_1^1(x_2)) \to (\forall x_1)(\forall x_2)A_2^2(x_1, x_2)).$$

Again we follow the method of proof of Proposition 4.28. First change the bound variables. It does not matter how we do it, except that the bound variables must be different from one another and from the free variables. We obtain (say)

$$(((\forall x_1)A_1^2(x_1, x_2) \to {\sim}(\exists x_3)A_1^1(x_3)) \to (\forall x_4)(\forall x_5)A_2^2(x_4, x_5)),$$

which is provably equivalent to the given *wf.* Now we proceed in steps, using different cases of the inductive proof above. First treat the quantifiers which are immediately preceded by ${\sim}$ (Case 1) to get

$$(((\forall x_1)A_1^2(x_1, x_2) \to (\forall x_3){\sim}A_1^1(x_3)) \to (\forall x_4)(\forall x_5)A_2^2(x_4, x_5)).$$

Now consider the parts of the form $(\mathscr{B} \to \mathscr{C})$ (Case 2). We obtain a sequence of *wfs.*, each one provably equivalent to the next, using the different parts of Proposition 4.25.

$$((\forall x_3)((\forall x_1)A_1^2(x_1, x_2) \to {\sim}A_1^1(x_3)) \to (\forall x_4)(\forall x_5)A_2^2(x_4, x_5)),$$

$$((\forall x_3)(\exists x_1)(A_1^2(x_1, x_2) \to {\sim}A_1^1(x_3)) \to (\forall x_4)(\forall x_5)A_2^2(x_4, x_5)),$$

$$(\exists x_3)(\forall x_1)((A_1^2(x_1, x_2) \to {\sim}A_1^1(x_3)) \to (\forall x_4)(\forall x_5)A_2^2(x_4, x_5)),$$

$$(\exists x_3)(\forall x_1)(\forall x_4)(\forall x_5)((A_1^2(x_1, x_2) \to {\sim}A_1^1(x_3)) \to A_2^2(x_4, x_5)).$$

This last *wf.* is in prenex form and is provably equivalent to the given *wf.*

▷ Note that this procedure does not lead to a unique answer. The order in which the quantifiers are brought out to the beginning is arbitrary. For example, the *wf.*

$$(\forall x_4)(\forall x_5)(\forall x_1)(\exists x_3)((A_1^2(x_1, x_2) \to {\sim}A_1^1(x_3)) \to A_2^2(x_4, x_5))$$

is also a possible answer to the above. However, the order of the quantifiers at the beginning of a prenex form formula does matter. It is only in particular cases that the order can be changed, and then usually only in particular ways, if the resulting *wf.* is to be provably equivalent to the original *wf.*

Prenex forms lead to a way of measuring the complexity of *wfs.* of K. It might seem at first that for a *wf.* in prenex form, the more quantifiers there are at the beginning the more complex will be its interpretation. However, consider the two *wfs.*

$$(\forall x_1)(\forall x_2)(\forall x_3)(\forall x_4)A_1^2(f_1^2(x_1, x_2), f_1^2(x_3, x_4)),$$

$$(\forall x_1)(\exists x_2)(\forall x_3)(\exists x_4)A_1^2(f_1^2(x_1, x_2), f_1^2(x_3, x_4)).$$

We can see that the former is much more easily interpreted. For example, consider the arithmetic interpretation of Chapter 3. The interpretations of these *wfs.* are respectively

for every $x, y, z, t \in D_N, x + y = z + t$,

and   for every $x \in D_N$ there is a $y \in D_N$ such that for every

$z \in D_N$ there is a $t \in D_N$ such that $x + y = z + t$.

The second is a much more complicated sentence, and it is difficult at first sight to see whether it is true or false. The complications arise because of the alternating quantifiers, and a measure of complexity is the number of alternations.

Definition 4.30

(i) Let $n > 0$. A *wf.* in prenex form is a Π_n-*form* if it starts with a universal quantifier and has $n - 1$ alternations of quantifiers.

(ii) Let $n > 0$. A *wf.* in prenex form is a Σ_n-*form* if it starts with an existential quantifier and has $n - 1$ alternations of quantifiers.

We shall not use these definitions, but they are important in the further study of the subject.

Example 4.31

(a) $(\forall x_1)(\forall x_2)(\exists x_3)A_1^3(x_1, x_2, x_3)$ is a Π_2-form.
(b) $(\forall x_1)(A_1^1(x_1) \rightarrow A_2^1(x_2))$ is a Π_1-form.
(c) $(\exists x_1)(\forall x_2)(\exists x_3)(\exists x_4)(A_1^2(x_1, x_2) \rightarrow A_1^2(x_3, x_4))$ is a Σ_3-form.

▷ There is another special form of *wf.* which has recently become prominent because of its use in logic programming via the language Prolog. This is what is called clausal form.

Definition 4.32

A *wf.* of $\mathscr{L}$ is in *clausal form* if it is a conjunction of clauses. A *clause* is a disjunction of *wfs.*, each of which is either an atomic formula or a negation of an atomic formula.

▷ This may become clearer after the following description of a procedure by which a clausal form may be obtained from any given *wf.* Let $\mathscr{A}$ be any *wf.* of $\mathscr{L}$.

(i)   Find a *wf.* $\mathscr{B}$ in prenex form which is provably equivalent to $\mathscr{A}$. (See Proposition 4.28.)

(ii)   Skolemise $\mathscr{B}$, i.e. eliminate all existential quantifiers, replacing occurrences of the quantified variables by terms involving Skolem functions or Skolem constants. This yields a *wf.* which may not be equivalent to $\mathscr{A}$, but which is weakly equivalent, in the sense given in Proposition 3.39.

(iii)   Delete all universal quantifiers. This gives a *wf.* $\mathscr{C}$ which may not be equivalent, but is true in the same interpretations as the *wf.* at the previous stage. (See Proposition 3.27.)

(iv)   The *wf.* $\mathscr{C}$ contains no quantifiers, so it is built up from atomic formulas using the other logical connectives. If we treat the atomic formulas as statement variables then $\mathscr{C}$ can be regarded as a statement form, as in propositional calculus. By Corollary 1.21, if $\mathscr{C}$ is not a tautology, then there is a *wf.* $\mathscr{D}$ in conjunctive normal form which is logically equivalent to $\mathscr{C}$.

(v)   Each conjunct in $\mathscr{D}$ is a disjunction of atomic formulas or negations of atomic formulas, i.e. a clause, so $\mathscr{D}$ is in clausal form.

It is important to realise that the steps above do not all produce logically equivalent formulas. The weakest step in this respect is step (ii), and the weak equivalence here provides the best we can say about the original *wf.* and the *wf.* in clausal form. (One is contradictory if and only if the other is contradictory.)

Proposition 4.33

Every *wf.* of $\mathscr{L}$ which is not a tautology is weakly equivalent to a *wf.* in clausal form.

Exercises

8   For each of the following formulas, find a formula in prenex form which is provably equivalent to it.

   (a) $(\forall x_1)A_1^1(x_1) \to (\forall x_2)A_1^2(x_1, x_2)$.

   (b) $(\forall x_1)(A_1^2(x_1, x_2) \to (\forall x_2)A_1^2(x_1, x_2))$.

   (c) $(\forall x_1)(A_1^1(x_1) \to A_1^2(x_1, x_2)) \to ((\exists x_2)A_1^1(x_2) \to (\exists x_3)A_1^2(x_2, x_3))$.

   (d) $(\exists x_1)A_1^2(x_1, x_2) \to (A_1^1(x_1) \to \sim(\exists x_3)A_1^2(x_1, x_3))$.

9   Let $\mathscr{A}(x_1)$ be a *wf.* in which x_2 does not occur, let $\mathscr{B}(x_2)$ be a *wf.* in which x_1 does not occur, and suppose that $\mathscr{A}$ and $\mathscr{B}$ contain no quantifiers. Show that the formula

$$((\exists x_1)\mathscr{A}(x_1) \to (\exists x_2)\mathscr{B}(x_2))$$

   is provably equivalent to formulas in prenex form of both Π_2 and Σ_2 forms.

10   Find a formula in Π_3 form which is provably equivalent to a formula in Σ_2 form.

11   For each of the *wfs.* in Exercise 8 above, find a *wf.* in clausal form which is weakly equivalent to it.

4.4 The Adequacy Theorem for *K*

We shall eventually prove the following proposition:

> If $\mathscr{A}$ is a logically valid *wf.* of $\mathscr{L}$, then $\mathscr{A}$ is a theorem of $K_{\mathscr{L}}$.

However, before we come to the proof, we need some preliminary work in order to be able to extend and apply the ideas used in the proof of the Adequacy Theorem for *L*.

First, the idea of extension can be easily generalised. (Here, as before, we shall write *K* instead of $K_{\mathscr{L}}$ unless we wish to emphasise the language being used.)

Definition 4.34

An *extension* of *K* is a formal system obtained by altering or enlarging the set of axioms so that all theorems of *K* remain theorems (and new theorems are possibly introduced). Similarly, given two extensions of *K*, one is an extension of the other if its class of theorems is larger (or, in the trivial case, the same).

Definition 4.35

A *first order system* is an extension of $K_{\mathscr{L}}$, for some first order language $\mathscr{L}$.

Definition 4.36

A first order system S is *consistent* if for no *wf.* $\mathcal{A}$ are both $\mathcal{A}$ and $(\sim\mathcal{A})$ theorems of S.

Proposition 4.37 (See Proposition 2.19)

Let S be a consistent first order system and let $\mathcal{A}$ be a *closed wf.* which is not a theorem of S. Then S^* is also consistent, where S^* is the extension of S obtained by including $(\sim\mathcal{A})$ as an additional axiom.

Proof. Suppose that S^* is inconsistent. Then for some *wf.* $\mathcal{B}$, $\vdash_{S^*} \mathcal{B}$ and $\vdash_{S^*} (\sim\mathcal{B})$. Now $\vdash_{S^*} (\sim\mathcal{B} \to (\mathcal{B} \to \mathcal{A}))$, by Proposition 4.3, since S^* is an extension of K. Hence $\vdash_{S^*} (\mathcal{B} \to \mathcal{A})$ by *MP*, and $\vdash_{S^*} \mathcal{A}$ by *MP* again.

There is therefore a proof of $\mathcal{A}$ in S^*. Such a proof is a deduction in S from $(\sim\mathcal{A})$. So we have

$$(\sim\mathcal{A}) \vdash_{S} \mathcal{A}.$$

Since $(\sim\mathcal{A})$ is closed, we can apply the Deduction Theorem to obtain

$$\vdash_{S} ((\sim\mathcal{A}) \to \mathcal{A}).$$

But

$$\vdash_{S} (((\sim\mathcal{A}) \to \mathcal{A}) \to \mathcal{A}), \text{ by Proposition 4.3.}$$

Therefore $\vdash_{S} \mathcal{A}$ by *MP*.

This contradicts the hypothesis that $\mathcal{A}$ is not a theorem of S, and so S^* must be consistent.

▷ Note that this proposition is an analogue of Proposition 2.19, but that we must impose the condition that $\mathcal{A}$ be a closed *wf.* In our use of this result this will be no handicap.

Definition 4.38

A first order system S is *complete* if for each closed *wf.* $\mathcal{A}$, either $\vdash_{S} \mathcal{A}$ or $\vdash_{S} (\sim\mathcal{A})$.

Notice that K is *not* complete. For example, the *wf.* $(\forall x_1)A_1^1(x_1)$ is not a theorem of K, nor is its negation.

Proposition 4.39 (See Proposition 2.21)

Let S be a consistent first order system. Then there is a consistent extension of S which is complete.

Proof. The proof follows precisely the same pattern as the proof of Proposition 2.21. Let $\mathscr{A}_0, \mathscr{A}_1, \ldots$ be an enumeration of all the closed wfs. of $\mathscr{L}$. We construct a sequence $S_0, S_1, \ldots$ of extensions of K as follows. Let S_0 be S. For $n > 0$, if $\underset{S_{n-1}}{\vdash} \mathscr{A}_{n-1}$, let S_n be the same as S_{n-1}, and if not $\underset{S_{n-1}}{\vdash} \mathscr{A}_{n-1}$, let S_n be the extension of S_{n-1} obtained by including $(\sim \mathscr{A}_{n-1})$ as an additional axiom. By Proposition 4.37, it is clear that each S_n is a consistent extension of K. Let S_∞ be the first order system which has as its axioms all the wfs. which are axioms of at least one of the S_n. Precisely as in the proof of Proposition 2.21 we can show that S_∞ is consistent and complete.

▷ It is at this point that our methods must become substantially different from those of Chapter 2. If we were to 'translate' Proposition 2.22 into the terminology of the predicate calculus, what we might obtain is: If S is a consistent first order system, then there is an interpretation of $\mathscr{L}$ in which every theorem of S is true. We shall indeed prove this, but the proof is rather difficult and it involves new ideas.

Up till now, the language $\mathscr{L}$ has been fixed, though arbitrary. In the next proof we shall have occasion to enlarge the language by adding an infinite list of new individual constants $b_0, b_1, \ldots$. This will certainly have the effect of introducing new wfs., new axioms and new theorems to K (for example, the new wf. $((\forall x_1)A_1^1(x_1) \to A_1^1(b_1))$ will be an axiom of the system with the enlarged language). However, if S is a consistent extension of $K_{\mathscr{L}}$, then the new system S^+ obtained as above by enlarging the language is also consistent. For if both $\mathscr{A}$ and $(\sim \mathscr{A})$ were theorems of S^+, then their proofs, being finite sequences of wfs., would involve only a finite number of the symbols $b_0, b_1, b_2, \ldots$. These proofs could then be converted into proofs in S by substituting for each of these symbols a variable which does not occur elsewhere in the proofs. We would thus obtain proofs in S of a wf. of $\mathscr{L}$ and its negation, which is impossible.

Proposition 4.40

Let S be a consistent extension of $K_{\mathscr{L}}$. Then there is an interpretation of $\mathscr{L}$ in which every theorem of S is true.

Proof. Enlarge the language $\mathscr{L}$ by including a sequence $b_0, b_1, b_2, \ldots$ of new individual constants. Denote by $\mathscr{L}^+$ the new language, and by S^+ and K^+ the new systems obtained from S and $K_{\mathscr{L}}$ respectively. Then S^+ is consistent, as above. Starting with S^+, let us define a sequence of first order systems $S_0, S_1, \ldots$ as follows. First enumerate in a list all the *wfs.* of $\mathscr{L}^+$ which contain just one free variable:

$$\mathscr{F}_0(x_{i_0}), \mathscr{F}_1(x_{i_1}), \ldots, \text{say.}$$

Of course, the $x_{i_0}, x_{i_1}, \ldots$ will not all be distinct. Now choose a subsequence $c_0, c_1, c_2, \ldots$ of the sequence $b_0, b_1, b_2, \ldots$ so that

(1) c_0 does not occur in $\mathscr{F}_0(x_{i_0})$,

and

(2)   for $n > 0$, $c_n \notin \{c_0, \ldots, c_{n-1}\}$ and c_n does not occur in any of $\mathscr{F}_0(x_{i_0}), \ldots, \mathscr{F}_n(x_{i_n})$.

We can do this since each *wf.* can contain occurrences of only finitely many (if any) of the b_i.

For each k, denote by $\mathscr{G}_k$ the *wf.*

$$(\sim(\forall x_{i_k})\mathscr{F}_k(x_{i_k}) \to \sim\mathscr{F}_k(c_k)).$$

Now let S_0 be S^+. Let S_1 be the extension of S_0 obtained by including $\mathscr{G}_0$ as a new axiom. For each $n > 1$, let S_n be the extension of S_{n-1} obtained by including $\mathscr{G}_{n-1}$ as a new axiom. Our procedure will be to show that each S_n is consistent, to obtain a consistent S_∞ from the sequence, as before, and then to apply Proposition 4.39 to obtain a consistent complete extension of S_∞. This will enable us to construct our interpretation as required.

S_0 is consistent. Let $n > 0$ and suppose that S_n is consistent, but S_{n+1} is not. Then there is a *wf.* $\mathscr{A}$ of $\mathscr{L}^+$ such that

$$\vdash_{S_{n+1}} \mathscr{A} \quad \text{and} \quad \vdash_{S_{n+1}} (\sim\mathscr{A}),$$

Now $\vdash_{S_{n+1}} (\mathscr{A} \to (\sim\mathscr{A} \to \sim\mathscr{B}))$, since $(\mathscr{A} \to (\sim\mathscr{A} \to \sim\mathscr{B}))$ is a tautology, for any *wf.* $\mathscr{B}$, so by two applications of *MP*, we obtain

$$\vdash_{S_{n+1}} (\sim\mathscr{B}), \text{ for any } wf. \ \mathscr{B}.$$

In particular,

$$\vdash_{S_{n+1}} (\sim\mathscr{G}_n).$$

Now a proof in S_{n+1} is just a deduction in S_n from $\mathscr{G}_n$, so we have

$$\mathscr{G}_n \vdash_{S_n} (\sim\mathscr{G}_n)$$

$\mathscr{G}_n$ is closed, so by the Deduction Theorem,

$$\underset{S_n}{\vdash} (\mathscr{G}_n \rightarrow (\sim \mathscr{G}_n)).$$

It follows now, as we have seen before, that

$$\underset{S_n}{\vdash} (\sim \mathscr{G}_n),$$

i.e.

$$\underset{S_n}{\vdash} \sim (\sim (\forall x_{i_n}) \mathscr{F}_n(x_{i_n}) \rightarrow \sim \mathscr{F}_n(c_n)).$$

But

$$\underset{S_n}{\vdash} (\sim (\sim (\forall x_{i_n}) \mathscr{F}_n(x_{i_n}) \rightarrow \sim \mathscr{F}_n(c_n)) \rightarrow \sim (\forall x_{i_n}) \mathscr{F}_n(x_{i_n}))$$

and

$$\underset{S_n}{\vdash} (\sim (\sim (\forall x_{i_n}) \mathscr{F}_n(x_{i_n}) \rightarrow \sim \mathscr{F}_n(c_n)) \rightarrow \mathscr{F}_n(c_n)),$$

since both of these wfs. are instances of tautologies. Hence, by *MP*, we have

$$\underset{S_n}{\vdash} \sim (\forall x_{i_n}) \mathscr{F}_n(x_{i_n}) \quad \text{and} \quad \underset{S_n}{\vdash} \mathscr{F}_n(c_n).$$

In the proof of $\mathscr{F}_n(c_n)$, replace each occurrence of c_n by y, where y is a variable not occurring elsewhere in the proof. What we obtain is a proof in S_n of $\mathscr{F}_n(y)$, since c_n does not occur in any of the axioms $\mathscr{G}_0, \mathscr{G}_1, \ldots, \mathscr{G}_{n-1}$ from which $\mathscr{F}_n(c_n)$ was derived in S_n. Thus

$$\underset{S_n}{\vdash} \mathscr{F}_n(y).$$

Therefore

$$\underset{S_n}{\vdash} (\forall y) \mathscr{F}_n(y) \text{ by Generalisation,}$$

and so

$$\underset{S_n}{\vdash} (\forall x_{i_n}) \mathscr{F}_n(x_{i_n}), \text{ by Proposition 4.18.}$$

We now have contradicted the consistency of S_n, so we must have: if S_n is consistent then so is S_{n+1}, for all $n \geq 0$. By induction, therefore, S_n is consistent for all n.

Let S_∞ be the system obtained by including as axioms all wfs. of $\mathscr{L}^+$ which are axioms of at least one of the S_n. S_∞ is consistent, for if it were not, then a contradiction could be derived using only finitely many of its axioms, and so could be derived in S_n for some n.

By Proposition 4.39, then, let T be an extension of S_∞ which is consistent and complete.

The way that our interpretation is now constructed is something new and may cause confusion. We have been concerned before with interpretations whose domains have consisted of mathematical objects, for example natural numbers or integers. According to the definition, however, the domain is required to be just a non-empty set. We define an interpretation I of $\mathscr{L}^+$ as follows:

(a) The domain D_I is the set of all *closed terms* of $\mathscr{L}^+$, i.e. terms not containing any variables (i.e., all individual constants and all terms built up from them using the function letters).

(b) The individual constants are their own interpretations.

(c) For $d_1, \ldots, d_n \in D_I$, $\bar{A}_i^n(d_1, \ldots, d_n)$ holds if $\vdash_T A_i^n(d_1, \ldots, d_n)$ and does not hold if $\vdash_T (\sim A_i^n(d_1, \ldots, d_n))$. This makes sense because T is complete, and $A_i^n(d_1, \ldots, d_n)$ is a closed *wf*.

(d) For $d_1, \ldots, d_n \in D_I$, $\bar{f}_i^n(d_1, \ldots, d_n)$ is given the value $f_i^n(d_1, \ldots, d_n)$. Note that since $d_1, \ldots, d_n$ are closed terms, so also is $f_i^n(d_1, \ldots, d_n)$. This defines the interpretation I. We must now show that every theorem of S is true in I.

Lemma. For any closed *wf.* $\mathscr{A}$ of $\mathscr{L}^+$, $\vdash_T \mathscr{A}$ if and only if $I \vDash \mathscr{A}$.

Proof. The proof is by induction on the number of connectives and quantifiers in $\mathscr{A}$.

Base step: $\mathscr{A}$ is an atomic formula, say $A_i^n(d_1, \ldots, d_n)$, where $d_1, \ldots, d_n$ are necessarily closed terms.

If $\vdash_T \mathscr{A}$ then $\vdash_T A_i^n(d_1, \ldots, d_n)$, so $\bar{A}_i^n(d_1, \ldots, d_n)$ holds in I, i.e. $I \vDash \mathscr{A}$. Similarly, if $I \vDash \mathscr{A}$, we can deduce $\vdash_T \mathscr{A}$.

Induction step: Let $\mathscr{A}$ not be an atomic formula, and suppose that the result holds for every *wf.* which is shorter than $\mathscr{A}$.

Case 1: $\mathscr{A}$ is $(\sim \mathscr{B})$. If $\vdash_T \mathscr{A}$ then $\vdash_T (\sim \mathscr{B})$. It follows that $\mathscr{B}$ is not a theorem of T, since T is consistent, and so $\mathscr{B}$ is not true in I, by the induction hypothesis. Hence $(\sim \mathscr{B})$ is true in I, since $\mathscr{B}$ is closed, i.e. $I \vDash \mathscr{A}$. Conversely, if $I \vDash \mathscr{A}$, then $I \vDash (\sim \mathscr{B})$, and so $\mathscr{B}$ is not true in I, and so $\mathscr{B}$ is not a theorem of T, by the induction hypothesis. Since T is complete, then, $(\sim \mathscr{B})$ is a theorem of T, i.e. $\vdash_T \mathscr{A}$.

Case 2: $\mathscr{A}$ is $(\mathscr{B} \to \mathscr{C})$. Suppose that $\mathscr{A}$ is not true in I. Then $\mathscr{B}$ is true and $\mathscr{C}$ is false. By the induction hypothesis, then, $\vdash_T \mathscr{B}$ and not $\vdash_T \mathscr{C}$. Since T is complete, we have $\vdash_T \mathscr{B}$ and $\vdash_T (\sim \mathscr{C})$. Now $\vdash_T (\mathscr{B} \to ((\sim \mathscr{C}) \to$

$\sim(\mathscr{B} \to \mathscr{C})))$ since this *wf.* is an instance of a tautology, so by *MP* used twice, we obtain

$$\underset{T}{\vdash} \sim(\mathscr{B} \to \mathscr{C}), \quad \text{i.e.} \underset{T}{\vdash} (\sim\mathscr{A}).$$

Since T is consistent, then, $\mathscr{A}$ is not a theorem of T.

Conversely, suppose that $\mathscr{A}$ is not a theorem of T. Then $\underset{T}{\vdash}(\sim\mathscr{A})$, by completeness of T, and so $\underset{T}{\vdash}\sim(\mathscr{B} \to \mathscr{C})$. But $(\sim(\mathscr{B} \to \mathscr{C}) \to \mathscr{B})$ and $(\sim(\mathscr{B} \to \mathscr{C}) \to \sim\mathscr{C})$ are tautologies, so we have

$$\underset{T}{\vdash} \mathscr{B} \quad \text{and} \quad \underset{T}{\vdash}(\sim\mathscr{C}),$$

so

$$\underset{T}{\vdash} \mathscr{B} \quad \text{and not} \quad \underset{T}{\vdash} \mathscr{C}, \quad \text{since } T \text{ is consistent.}$$

By the induction hypothesis, then,

$$I \vDash \mathscr{B} \quad \text{and not} \quad I \vDash \mathscr{C}.$$

So $\mathscr{B}$ is true in I and $\mathscr{C}$ is false in I. By Remark 3.25 (d), then, $(\mathscr{B} \to \mathscr{C})$ is false in I, and therefore not true in I.

Case 3: $\mathscr{A}$ is $(\forall x_i)\mathscr{B}(x_i)$. First, if x_i does not occur free in $\mathscr{B}$, then $\mathscr{B}$ is closed, so that, by the induction hypothesis, $\underset{T}{\vdash} \mathscr{B}$ if and only if $I \vDash \mathscr{B}$. Also we know that $\underset{T}{\vdash} \mathscr{B}$ if and only if $\underset{T}{\vdash} (\forall x_i)\mathscr{B}$, and that $I \vDash \mathscr{B}$ if and only if $I \vDash (\forall x_i)\mathscr{B}$. Hence in this case, $\underset{T}{\vdash} \mathscr{A}$ if and only if $I \vDash \mathscr{A}$.

Second, if x_i does occur free in $\mathscr{B}(x_i)$, then it is the only free variable in $\mathscr{B}(x_i)$, since $\mathscr{A}$ is closed. So $\mathscr{B}(x_i)$ is one of the *wfs.* in the sequence $\mathscr{F}_0(x_{i_0}), \mathscr{F}_1(x_{i_1}), \ldots$, say $\mathscr{B}(x_i)$ is $\mathscr{F}_m(x_{i_m})$. Then $\mathscr{A}$ is $(\forall x_{i_m})\mathscr{F}_m(x_{i_m})$. Suppose that $I \vDash \mathscr{A}$. By Proposition 4.4, we have (from axiom $(K5)$)

$$I \vDash ((\forall x_{i_m})\mathscr{F}_m(x_{i_m}) \to \mathscr{F}_m(c_m)).$$

Hence $I \vDash \mathscr{F}_m(c_m)$. Now $\mathscr{F}_m(c_m)$ has fewer connectives and quantifiers than $\mathscr{A}$, so by the induction hypothesis, $\underset{T}{\vdash} \mathscr{F}_m(c_m)$. We wish to show that $\underset{T}{\vdash} \mathscr{A}$, so suppose the contrary, i.e. $\underset{T}{\vdash} (\sim\mathscr{A})$, since T is complete. i.e.

$$\underset{T}{\vdash} \sim(\forall x_{i_m})\mathscr{F}_m(x_{i_m}).$$

But

$$\underset{T}{\vdash} (\sim(\forall x_{i_m})\mathscr{F}_m(x_{i_m}) \to \sim\mathscr{F}_m(c_m)),$$

since $\mathscr{G}_m$ is an axiom of T. Therefore, by *MP*,

$$\underset{T}{\vdash} (\sim\mathscr{F}_m(c_m)).$$

This contradicts the consistency of T, so we must have $\underset{T}{\vdash} \mathscr{A}$, as required.

Conversely, let $\vdash_T \mathscr{A}$, and suppose that $\mathscr{A}$ is not true in I, i.e. not $I \vDash (\forall x_{i_m}) \mathscr{F}_m(x_{i_m})$. Then there is an element d of D_I such that $I \vDash (\sim\mathscr{F}_m(d))$. To see this, observe that there is a valuation in I which does not satisfy $(\forall x_{i_m}) \mathscr{F}_m(x_{i_m})$, and so there is a valuation v which does not satisfy $\mathscr{F}_m(x_{i_m})$. Now $v(x_{i_m}) \in D_I$, i.e. $v(x_{i_m})$ is a closed term, d, say, and such a term is necessarily free for x_{i_m} in $\mathscr{F}_m(x_{i_m})$. Also $v(d) = d$. Thus $v(x_{i_m}) = v(d)$. By the result of Exercise 3.23, then, v does not satisfy $\mathscr{F}_m(d)$, and so $\mathscr{F}_m(d)$ is not true in I. Hence $I \vDash (\sim F_m(d))$, as required. But $\vdash_T (\forall x_{i_m}) \mathscr{F}_m(x_{i_m})$, so $\vdash_T \mathscr{F}_m(d)$, by axiom $(K5)$ and *MP*. By the induction hypothesis, then, $I \vDash \mathscr{F}_m(d)$. But $\mathscr{F}_m(d)$ and $\sim\mathscr{F}_m(d)$ cannot both be true in I, so $\vdash_T \mathscr{A}$ implies $I \vDash \mathscr{A}$ in this case.

This completes the inductive proof of the lemma, so now we know that every theorem of T is true in the interpretation I. Every theorem of S is a theorem of T, since T was obtained from S merely by enlarging the language and adding new axioms. Thus every *wf.* of $\mathscr{L}^+$ which is a theorem of S is true in I. Of course every theorem of S is a *wf.* of $\mathscr{L}$ and I contains interpretations of *wfs.* other than those of $\mathscr{L}$, so let us restrict I by excluding the interpretations of the individual constants $b_0, b_1, \ldots$ and terms depending on them, but leaving D_I unchanged. This gives us an interpretation of $\mathscr{L}$, and every theorem of S is true in this interpretation.

$\triangleright$ We still have not proved the Adequacy Theorem, but the considerable effort required in proving Proposition 4.40 above will enable us to complete the proof of the Adequacy Theorem quite easily.

Proposition 4.41 (The Adequacy Theorem for $K_{\mathscr{L}}$)

If $\mathscr{A}$ is a logically valid *wf.* of $\mathscr{L}$, then $\mathscr{A}$ is a theorem of $K_{\mathscr{L}}$.

Proof. Let $\mathscr{A}$ be a logically valid *wf.* of $\mathscr{L}$ and let $\mathscr{A}'$ be its universal closure. It follows from Corollary 3.28 that $\mathscr{A}'$ must be logically valid also. Suppose that $\mathscr{A}$ is not a theorem of $K_{\mathscr{L}}$. Then by Proposition 4.19, $\mathscr{A}'$ is not a theorem of $K_{\mathscr{L}}$. If we include $\sim\mathscr{A}'$ as an additional axiom, we obtain a new system $K'_{\mathscr{L}}$, which is consistent, by Proposition 4.37. Hence, by Proposition 4.40, there is an interpretation of $\mathscr{L}$ in which each theorem of $K'_{\mathscr{L}}$ is true. In particular, $\sim\mathscr{A}'$ is true in this interpretation, and so $\mathscr{A}'$ is false ($\mathscr{A}'$ is necessarily closed). This contradicts the logical validity of $\mathscr{A}'$, and so $\mathscr{A}$ must be a theorem of $K_{\mathscr{L}}$.

$\triangleright$ Our present task is thus completed. We have shown that the theorems of $K_{\mathscr{L}}$ are just the logically valid *wfs.* of $\mathscr{L}$. Though the proof of this has caused us some difficulty (it was first proved by Gödel in 1930, and the proof we have given is a later proof due essentially to Henkin), the result is no surprise. The system $K_{\mathscr{L}}$ was constructed in the way that it was just in order that one might be able to prove everything that might be expected to be provable by the methods of ordinary or intuitive logic, namely 'logical truths' (i.e. all logically valid *wfs.*). Nevertheless, the Adequacy Theorem is a central one, confirming as it does that $K_{\mathscr{L}}$ does what is expected of it.

Exercises

12   Show that an extension S of $K_{\mathscr{L}}$ is inconsistent if and only if every *wf.* of $\mathscr{L}$ is a theorem of S.

13   Let S be a consistent first order system such that, for every closed *wf.* $\mathscr{A}$ of S, if the system obtained by including $\mathscr{A}$ as an additional axiom is consistent then $\mathscr{A}$ is a theorem of S. Prove that S is complete.

14   Let $\mathscr{A}$ and $\mathscr{B}$ be *wfs.* of $\mathscr{L}$ such that $(\mathscr{A} \vee \mathscr{B})$ is a theorem of $K_{\mathscr{L}}$. Is it necessarily the case that either $\mathscr{A}$ or $\mathscr{B}$ is a theorem of $K_{\mathscr{L}}$?

15   Let $\mathscr{L}$ be a first order language with infinitely many predicate letters. Show that $K_{\mathscr{L}}$ has infinitely many different consistent extensions.

4.5 Models

Proposition 4.41 has many consequences, and we shall mention some of them here. In order to do so it is convenient to introduce a new notion at this point, that of a model.

Definition 4.42

(i) Let Γ be a set of *wfs.* of $\mathscr{L}$. An interpretation of $\mathscr{L}$ in which each element of Γ is true is called a *model* of Γ.

(ii) If S is a first order system, a *model* of S is an interpretation in which every theorem of S is true.

Proposition 4.43

Let S be a first order system, and let I be an interpretation in which every axiom of S is true. Then I is a model of S.

Proof. (Cf. the proof of Proposition 4.5)

Suppose that I is an interpretation in which every axiom of S is true. Let $\mathscr{A}$ be a theorem of S. We show that $\mathscr{A}$ must be true in I by induction on the number n of *wfs.* in a proof of $\mathscr{A}$.

Base step: $n = 1$. $\mathscr{A}$ is an axiom of S, and therefore true in I.

Induction step: $n > 1$. Suppose that every theorem with a shorter proof is true in I.

Case 1: $\mathscr{A}$ follows from previous wfs. in the proof by MP, say from $\mathscr{B}$ and $(\mathscr{B} \to \mathscr{A})$. By the induction hypothesis, $\mathscr{B}$ and $(\mathscr{B} \to \mathscr{A})$ are true in I, so by Proposition 3.26, $\mathscr{A}$ is true in I.

Case 2: $\mathscr{A}$ follows from a previous wf. in the proof by Generalisation. Then $\mathscr{A}$ is $(\forall x_i)\mathscr{B}$, say, and the previous wf. is $\mathscr{B}$. By the induction hypothesis, $\mathscr{B}$ is true in I, so by Proposition 3.27, $(\forall x_i)\mathscr{B}$ is true in I.

Case 3: $\mathscr{A}$ is an axiom of S. As above, $\mathscr{A}$ must be true in I.

This completes our inductive proof that every theorem of S is true in I. It follows that I is a model of S.

$\triangleright$ What we have shown is that a model, under Definition 4.42(ii), of a first order system S, is the same thing as a model, under Definition 4.42(i), of the set of axioms of S.

Note that it is a trivial consequence of Proposition 4.5 that any interpretation of $\mathscr{L}$ is a model of $K_{\mathscr{L}}$, since every theorem of $K_{\mathscr{L}}$ is true in every interpretation. The notion of model is more important in the context of extensions of $K_{\mathscr{L}}$, in which the class of theorems is larger and for which there are interpretations which are not models.

Proposition 4.40 can now be restated as: if a first order system S is consistent, then it has a model. In fact we have the following proposition.

Proposition 4.44

A first order system S is consistent if and only if it has a model.

Proof. The implication one way has been proved. Suppose that S has a model, say I, and that S is inconsistent. Then $\vdash_S \mathscr{A}$ and $\vdash_S (\sim\mathscr{A})$, for some wf. $\mathscr{A}$. Now all theorems of S are true in the model I, so $\mathscr{A}$ and $(\sim\mathscr{A})$ are both true in I. This is impossible, by Remark 3.25(b), so S must be consistent.

Example 4.45

Let $\mathscr{A}$ be a closed wf. of $\mathscr{L}$ such that neither $\mathscr{A}$ nor $(\sim\mathscr{A})$ are theorems of K. Then, by Proposition 4.37, the systems K^1 and K^2, obtained by including as a new axiom $\mathscr{A}$ and $(\sim\mathscr{A})$ respectively, are both consistent. Hence K^1 and K^2 each has a model. The model for K^1 is an interpretation I_1 in which $\mathscr{A}$ is true. The model for K^2 is an interpretation I_2 in which $(\sim\mathscr{A})$ is true. Thus I_1 cannot be a model for K^2 and likewise I_2 cannot be a model for K^1. It follows from this that any consistent first order system which is not complete (i.e. in which there is a closed *wf.* $\mathscr{A}$ such that neither $\mathscr{A}$ nor $(\sim\mathscr{A})$ are theorems) has at least two models which are essentially different.

▷ We must be careful to avoid falling into the trap of supposing that because a particular *wf.* is true in a model of a system S, it must be a theorem of S. The above example shows how this need not hold. However we can do something in this direction, as a consequence of Proposition 4.44.

Proposition 4.46

Let S be a consistent first order system, and let $\mathscr{A}$ be a closed *wf.* which is true in *every* model of S. Then $\mathscr{A}$ is a theorem of S.

Proof. Let $\mathscr{A}$ be a closed *wf.* which is true in every model of S, and suppose that $\mathscr{A}$ is not a theorem of S. Then, by Proposition 4.37, the system S' obtained from S by including $(\sim\mathscr{A})$ as an additional axiom is consistent. S' therefore has a model, M, say, by Proposition 4.44. $(\sim\mathscr{A})$ is true in M, and so $\mathscr{A}$ is false in M. But M is a model of S (since S' is an extension of S). This contradicts the hypothesis that $\mathscr{A}$ is true in every model of S, and so $\mathscr{A}$ must be a theorem of S.

Proposition 4.47 (Löwenheim–Skolem Theorem)

If a first order system S has a model, then S has a model whose domain is a countable set (a set is countable if its elements may be put in one–one correspondence with the set of natural numbers).

Proof. If S has a model, then S is consistent, by Proposition 4.44. If S is consistent then the proof of Proposition 4.40 shows that S has a model of a particular nature. Its domain consists of the set of closed terms in an enlarged language. This set is countable. It is shown to be countable by describing a procedure for writing out a list (infinite, of course) which eventually includes each closed term. This can be done in several ways, and it is left as an exercise.

▷ This proposition has some surprising consequences, as we shall see in a later chapter.

Proposition 4.48 (The Compactness Theorem)

If each finite subset of the set of axioms of a first order system S has a model, then S itself has a model.

Proof. Suppose that each finite set of axioms of S has a model, but that S does not have a model. Then S is inconsistent, by Proposition 4.44. Hence $\vdash_S \mathscr{A}$ and $\vdash_S (\sim\mathscr{A})$, for some *wf.* $\mathscr{A}$. But these proofs can contain instances of only finitely many of the axioms of S. Let Γ be the set of all axioms of S which are used in these proofs. Γ is finite and therefore has a model. Therefore there is an interpretation I in which every member of Γ is true. It follows that $\mathscr{A}$ and $(\sim\mathscr{A})$ must both be true in I. This is because the rules of deduction, *MP* and Generalisation, preserve truth in an interpretation (see the proof of Proposition 4.43). But $\mathscr{A}$ and $(\sim\mathscr{A})$ cannot both be true in I, so we have the necessary contradiction and thus S must have a model.

▷ This proposition is sometimes stated in a slightly different form, which we give as a corollary.

Corollary 4.49

Let Γ be an infinite set of *wfs.* of $K_{\mathscr{L}}$. Then Γ has a model if each finite subset of Γ has a model.

▷ There is a way in which models give rise to first order systems. Let S be a first order system, and suppose that S is consistent. Then S has a model, I, say. Let us suppose that S is not complete, so that for some closed *wf.* $\mathscr{A}$, neither $\mathscr{A}$ nor $(\sim\mathscr{A})$ are theorems of S. However, in I this *wf.* $\mathscr{A}$ is either true or false, i.e. either $I \vDash \mathscr{A}$ or $I \vDash (\sim\mathscr{A})$. The model I similarly gives a 'truth value' to each closed *wf.* Let us define a new first order system $S(I)$ by including as axioms all *wfs.* which are true in I. Then the theorems of $S(I)$ are all axioms of $S(I)$, since any consequence of *wfs.* which are true in I will also be true in I. $S(I)$ is consistent, for if $\vdash_{S(I)} \mathscr{A}$ and $\vdash_{S(I)} (\sim\mathscr{A})$ then $\mathscr{A}$ and $(\sim\mathscr{A})$ are both true in I, which is impossible. Note that I is a model for $S(I)$. Also $S(I)$ is complete, for, given any closed *wf.* $\mathscr{A}$, either $I \vDash \mathscr{A}$ or $I \vDash (\sim\mathscr{A})$, so certainly we have either $\vdash_{S(I)} \mathscr{A}$ or $\vdash_{S(I)} (\sim\mathscr{A})$.

Exercises

16   Let Γ be a set of *wfs.* of $\mathscr{L}$ and let M be a model of Γ. Show that if $\Gamma \vdash_{K_{\mathscr{L}}} \mathscr{A}$, then $\mathscr{A}$ is true in M. Does the converse also hold?

17   Let S be a consistent complete extension of $K_{\mathscr{L}}$. Prove that any two models of S are elementarily equivalent, i.e. every closed *wf.* which is true in one model is true in the other.

18   Let S be a consistent extension of $K_{\mathscr{L}}$, and let M be a model of S. Define an extension S^+ of S as follows: include as additional axioms all atomic

formulas of $\mathscr{L}$ which are true in M. Prove that S^+ is consistent. Is S^+ necessarily complete?

19 Let S be a consistent extension of $K_{\mathscr{L}}$, and let M be a model of S. Define an extension $\hat{S}$ of S as follows: include as additional axioms all *closed* atomic formulas of $\mathscr{L}$ which are true in M and the negations of all *closed* atomic formulas of $\mathscr{L}$ which are not true in M. Prove that $\hat{S}$ is consistent. Is $\hat{S}$ necessarily complete?

20 Let S be a consistent extension of $K_{\mathscr{L}}$, where $\mathscr{L}$ is the first order language containing variables, individual constants $a_1, a_2, \ldots$, only one predicate letter A_1^1, and no function letters. An interpretation I of $\mathscr{L}$ can be thought of as a set D_I with a distinguished subset A_I consisting of all those $x \in D_I$ such that $\bar{A}_1^1(x)$ holds in I. Suppose that for each $n \geq 1$ there is a model M_n of S in which $\bar{a}_i \in A_{M_n}$ for $1 \leq i \leq n$. Prove that there is a model M of S in which $\bar{a}_i \in A_M$ for every i.

5
Mathematical systems

5.1 Introduction

Chapters 1 to 4 are not mathematics. The systems L and $K_{\mathscr{L}}$ are systems of logical deduction. We have had to use some mathematical techniques in order to obtain proofs of propositions, but those techniques have been of an elementary nature, principally properties of natural numbers. The mathematician interested in the foundations of his subject seeks to clarify the assumptions he makes and the procedures he uses. We can use the system $K_{\mathscr{L}}$ in such a clarification. $K_{\mathscr{L}}$ embodies procedures of logical deduction as used by mathematicians. We have seen that the absence of restrictions on the language $\mathscr{L}$ make our results about $K_{\mathscr{L}}$ very general, and that the symbols of a given $\mathscr{L}$ can be interpreted in many different ways. For any $\mathscr{L}$, however, there is a class of wfs. whose truth does not depend on the interpretation of the symbols, namely the class of logically valid wfs., i.e. the class of theorems of $K_{\mathscr{L}}$. If $\mathscr{L}$ is interpreted in a mathematical way, as it is in our examples, the theorems of $K_{\mathscr{L}}$ are interpreted as mathematical truths. They are mathematical statements which are true because of their logical structure rather than because of their mathematical content. For example, in the arithmetic interpretation N, the wf.

$$(\forall x_1)(\forall x_2)(A_1^2(x_1, x_2) \rightarrow A_1^2(x_1, x_2)),$$

which is logically valid, is interpreted as a mathematical statement, namely: 'for all natural numbers x and y, if $x = y$ then $x = y$', which is true by virtue of its logical structure. On the other hand, the wf.

$$(\forall x_1)(\forall x_2)(A_1^2(x_1, x_2) \rightarrow A_1^2(x_2, x_1))$$

is interpreted as the mathematical statement: 'for all natural numbers x and y, if $x = y$ then $y = x$', which is true. That it is true is, however, a consequence of the meaning of ' $=$ ' rather than merely its logical structure. Indeed this wf. is not logically valid. It is not difficult to find an interpretation in which A_1^2 is not interpreted as $=$, in which it is false. It follows that this wf. is not a theorem of $K_{\mathscr{L}}$. Thus the theorems of $K_{\mathscr{L}}$

have in themselves no mathematical value. Each of our mathematical formal systems will be an extension of some $K_{\mathscr{L}}$, obtained by including appropriate additional axioms so that the theorems of the system will represent mathematical truths as well as logical truths. If our formal system is to be a mathematical system then it is clearly desirable to have as theorems all *wfs.* whose interpretations are mathematical truths (or, if that is not possible, as many such *wfs.* as we can).

What constitutes a mathematical truth depends to a very large extent on the mathematical context. For example, the statement

$$(\forall x)(\forall y)(xy = yx)$$

is true when regarded as a statement about natural numbers, but is not necessarily true when regarded as a statement about elements of an arbitrary group. We shall show, by means of examples, how different mathematical contexts can be represented by different formal systems, so that, in particular, the above statement would be the interpretation of a theorem of formal arithmetic but the interpretation of a non-theorem of formal group theory. The context will determine the language $\mathscr{L}$ (as in the case of arithmetic) and it will also determine a set of *proper axioms*. The word 'proper' is used in order to distinguish these from $(K1)$–$(K6)$, which are *logical axioms*, and are common to all our systems. Having specified $\mathscr{L}$, the proper axioms are *wfs.* of $\mathscr{L}$ which, when added as new axioms, give an extension of $K_{\mathscr{L}}$ in which mathematical truths of the particular context (as well as logical truths) appear as interpretations of theorems.

5.2   First order systems with equality

Mathematics can very rarely do without the relation of equality. The symbol ' = ' does not appear in our formal languages, but we have used it in examples as the interpretation of the predicate symbol A_1^2. In all our examples of mathematical systems we shall include A_1^2 in the language, and = will be its intended interpretation. As we observed above, the *wf.* $(\forall x_1)(\forall x_2)(A_1^2(x_1, x_2) \to A_1^2(x_2, x_1))$ is not a theorem of $K_{\mathscr{L}}$, but we would like it to be a theorem of our mathematical extensions of $K_{\mathscr{L}}$. One way of ensuring this would be to include it among the proper axioms of each mathematical system. But there are clearly other *wfs.* which would require similar treatment, for example $(\forall x_1)A_1^2(x_1, x_1)$. We do not need to include as axioms all such *wfs.*, but we take as *axioms for equality* a set of them from which others may be deduced.

$(E7)$ $A_1^2(x_1, x_1)$.

(E8) $A_1^2(t_k, u) \to A_1^2(f_i^n(t_1, \ldots, t_k, \ldots, t_n), f_i^n(t_1, \ldots, u, \ldots, t_n))$,
where $t_1, \ldots, t_n$, u are any terms, and f_i^n is any function
letter of $\mathscr{L}$.

(E9) $(A_1^2(t_k, u) \to (A_i^n(t_1, \ldots, t_k, \ldots, t_n) \to A_i^n(t_1, \ldots, u, \ldots, t_n)))$,
where $t_1, \ldots, t_n$, u are any terms, and A_i^n is any predicate
symbol of $\mathscr{L}$.

Notes 5.1

(a) (E8) and (E9) are axiom schemes, each representing a number of
axioms, possibly infinitely many, depending on the number of function
letters and predicate symbols in $\mathscr{L}$.

(b) All of these axioms have free variables occurring in them. They
have been written in this way for the sake of clarity and for ease of
application later. We know, however, that for any wf. $\mathscr{A}$ whose
universal closure is $\mathscr{A}'$, $\mathscr{A} \underset{K_{\mathscr{L}}}{\vdash} \mathscr{A}'$ and $\mathscr{A}' \underset{K_{\mathscr{L}}}{\vdash} \mathscr{A}$, so an equivalent set of
axioms would be the universal closures of these.

(c) As a consequence of (b), and Proposition 4.18, regarding change
of bound variables, the fact that the variable x_1 in particular appears in
(E7) has no significance. For example, $A_1^2(x_5, x_5)$ is a consequence of
(E7), by means of the deduction:

(1) $A_1^2(x_1, x_1)$ (E7)

(2) $(\forall x_1)A_1^2(x_1, x_1)$ (1), Generalisation

(3) $(\forall x_5)A_1^2(x_5, x_5)$ (2), Proposition 4.18

(4) $(\forall x_5)A_1^2(x_5, x_5) \to A_1^2(x_5, x_5)$ (K5)

(5) $A_1^2(x_5, x_5)$ (3), (4), MP.

All mathematical systems which we describe will be extensions of $K_{\mathscr{L}}$
(for some $\mathscr{L}$) which include amongst their axioms (E7) and all appropriate (depending on $\mathscr{L}$) instances of (E8) and (E9).

Remark 5.2

The need to include (E7) should be clear. It ensures that in any model
the interpretation of A_1^2 behaves in one respect like $=$. (E8) and (E9)
are more complex, but their inclusion ensures that in any model, the
interpretation of A_1^2 behaves like $=$ in another respect, namely that
equals may be substituted for one another.

Definition 5.3

The axioms (E7), (E8) and (E9) are called *axioms for equality*. Any
extension of $K_{\mathscr{L}}$ which includes amongst its axioms (E7) and all appropriate instances of (E8) and (E9) is called a *first order system with
equality*.

Proposition 5.4

Let S be a first order system with equality. Then the following are theorems of S.

  (i) $(\forall x_1)A_1^2(x_1, x_1)$,

  (ii) $(\forall x_1)(\forall x_2)(A_1^2(x_1, x_2) \rightarrow A_1^2(x_2, x_1))$,

  (iii) $(\forall x_1)(\forall x_2)(\forall x_3)(A_1^2(x_1, x_2) \rightarrow (A_1^2(x_2, x_3) \rightarrow A_1^2(x_1, x_3)))$.

Proof. (i) Immediate, by Generalisation, from $(E7)$.
(ii) We give a proof in S.

  (1) $A_1^2(x_1, x_2) \rightarrow (A_1^2(x_1, x_1) \rightarrow A_1^2(x_2, x_1))$ $(E9)$

  (2) $(A_1^2(x_1, x_2) \rightarrow (A_1^2(x_1, x_1) \rightarrow A_1^2(x_2, x_1))) \rightarrow$
         $((A_1^2(x_1, x_2) \rightarrow A_1^2(x_1, x_1)) \rightarrow (A_1^2(x_1, x_2) \rightarrow$
         $A_1^2(x_2, x_1)))$ $(K2)$

  (3) $(A_1^2(x_1, x_2) \rightarrow A_1^2(x_1, x_1)) \rightarrow (A_1^2(x_1, x_2) \rightarrow A_1^2(x_2, x_1))$
                                               $(1), (2), MP$

  (4) $(A_1^2(x_1, x_1) \rightarrow (A_1^2(x_1, x_2) \rightarrow A_1^2(x_1, x_1)))$ $(K1)$

  (5) $A_1^2(x_1, x_1)$ $(E7)$

  (6) $(A_1^2(x_1, x_2) \rightarrow A_1^2(x_1, x_1))$ $(4), (5), MP$

  (7) $(A_1^2(x_1, x_2) \rightarrow A_1^2(x_2, x_1))$ $(3), (6), MP$

  (8) $(\forall x_1)(\forall x_2)(A_1^2(x_1, x_2) \rightarrow A_1^2(x_2, x_1))$ (7), Generalisation

(iii) Again we give a proof in S.

  (1) $(A_1^2(x_2, x_1) \rightarrow (A_1^2(x_2, x_3) \rightarrow A_1^2(x_1, x_3)))$ $(E9)$

  (2) $(A_1^2(x_1, x_2) \rightarrow A_1^2(x_2, x_1))$ (ii) above

  (3) $(A_1^2(x_1, x_2) \rightarrow (A_1^2(x_2, x_3) \rightarrow A_1^2(x_1, x_3)))$ $(1), (2), HS$
  (4) $(\forall x_1)(\forall x_2)(\forall x_3)(A_1^2(x_1, x_2) \rightarrow (A_1^2(x_2, x_3) \rightarrow A_1^2(x_1, x_3)))$
                                               (3), Generalisation

▷ Thus, since each of (i), (ii), (iii) in the above proposition must be true in any model of S, the symbol A_1^2 will be interpreted in any model by a relation which is reflexive, symmetric and transitive, i.e. an equivalence relation. Now $=$ is the intended interpretation for A_1^2. In an arbitrary interpretation, the axioms could well be false, so A_1^2 could be interpreted by any binary relation, but in a model of S we have seen that the axioms must be true and, as above, A_1^2 must be interpreted as an equivalence relation. However, the axioms $(E7)$, $(E8)$ and $(E9)$ do not ensure that in any model of S the interpretation of A_1^2 is actually $=$.

Example 5.5

Consider the first order language $\mathscr{L}$ with variables $x_1, x_2, \ldots$, function letter f_1^2, and predicate letter A_1^2. Define an interpretation I as follows. D_I is the set $\mathbb{Z}$ of all integers, $\bar{f}_1^2(x, y)$ is $x + y$, and $\bar{A}_1^2(x, y)$ holds if and only if $x \equiv y \pmod 2$, for $x, y \in \mathbb{Z}$. The axioms for equality are true in this interpretation.

For $(E7)$, its interpretation is $x \equiv x \pmod 2$, which is true.

For $(E8)$, as a particular case, consider

$$A_1^2(x_1, x_2) \to A_1^2(f_1^2(x_1, x_3), f_1^2(x_2, x_3)).$$

This is interpreted as:

$$\text{if } x \equiv y \pmod 2 \quad \text{then } x + z \equiv y + z \pmod 2$$

which is true. Verification of $(E8)$ in its full generality is left as an exercise.

For $(E9)$, there are only two instances to be verified, since L contains only one predicate letter. These are

$$(A_1^2(t, u) \to (A_1^2(t, v) \to A_1^2(u, v)))$$

and

$$(A_1^2(t, u) \to (A_1^2(v, t) \to A_1^2(v, u))).$$

The interpretations of these are, respectively,

$$\text{if } x \equiv y \pmod 2 \quad \text{then } x \equiv z \pmod 2 \quad \text{implies } y \equiv z \pmod 2$$

and

$$\text{if } x \equiv y \pmod 2 \quad \text{then } z \equiv x \pmod 2 \quad \text{implies } z \equiv y \pmod 2,$$

which are true.

▷ This example shows that in a model of $(E7)$, $(E8)$ and $(E9)$ the symbol A_1^2 need not necessarily be interpreted by $=$. However, the following proposition restores the situation.

Proposition 5.6

If S is a consistent first order system with equality, then S has a model in which the interpretation of A_1^2 is $=$.

Proof. By Proposition 4.44, if S is consistent, then S has a model, M, say. $\bar{A}_1^2$ is an equivalence relation on D_M because of Proposition 5.4.

Denote the equivalence class containing x by $[x]$. Now define a new interpretation M^* as follows. The domain of M^* is $\{[x]: x \in D_M\}$, a_i is interpreted by $[\bar{a}_i]$, for each i, f_i^n is interpreted by $\hat{f}_i^n$, where, for $y_1, \ldots, y_n \in D_M$,

$$\hat{f}_i^n([y_1], \ldots, [y_n]) = [\bar{f}_i^n(y_1, \ldots, y_n)],$$

and A_i^n is interpreted by $\hat{A}_i^n$, where, for $y_1, \ldots, y_n \in D_M$

$$\hat{A}_i^n([y_1], \ldots, [y_n])$$

holds if and only if $\bar{A}_i^n(y_1, \ldots, y_n)$ holds, where $\bar{a}_i$, $\bar{f}_i^n$, $\bar{A}_i^n$ are the interpretations of the symbols of $\mathscr{L}$ in M.

It is a lengthy but not difficult task to verify that these are well defined and that M^* is a model of S. For example, let f be a one-place function letter of $\mathscr{L}$ and $\bar{f}$ its interpretation in M. Suppose that a and b are members of D_M and that $[a] = [b]$. We have to show that $[\bar{f}(a)] = [\bar{f}(b)]$. Now

$$\underset{S}{\vdash} (A_1^2(x_1, x_2) \to A_1^2(f(x_1), f(x_2))). \tag{E8}$$

Hence $(A_1^2(x_1, x_2) \to A_1^2(f(x_1), f(x_2)))$ is true in M, since M is a model, and so $\bar{A}_1^2(a, b)$ implies $\bar{A}_1^2(f(a), f(b))$, i.e. $[a] = [b]$ implies $[\bar{f}(a)] = [\bar{f}(b)]$.

Also the interpretation of A_1^2 in M^* is $=$, since $\hat{A}_1^2([x], [y])$ holds if and only if $\bar{A}_1^2(x, y)$ holds, i.e. if and only if $[x] = [y]$.

▷ This proof can be well illustrated by our last example in which we gave a model where A_1^2 was not interpreted as $=$. In that example we had $\bar{A}_1^2(x, y)$ if and only if $x \equiv y \pmod 2$ (x and y integers). Define a new model, with domain $\{[0], [1]\}$, in which f_1^2 and A_1^2 are interpreted by $\hat{f}_1^2$ and $\hat{A}_1^2$, given by

$$\hat{f}_1^2([x], [y]) = [\bar{f}_1^2(x, y)] = [x + y],$$
$$\hat{A}_1^2([x], [y]) \text{ holds if and only if } \bar{A}_1^2(x, y) \text{ holds},$$

i.e.   if and only if $x \equiv y \pmod 2$
i.e.   if and only if $[x] = [y]$.

Definition 5.7

Let S be a first order system with equality. A *normal* model of S is a model in which A_1^2 is interpreted as $=$.

We shall be concerned in what follows mostly with normal models, since they represent the intended mathematical situation regarding the interpretation of A_1^2.

Note. Of course it does not matter that we have chosen A_1^2 to stand for equality. We could have chosen, say, A_{17}^2 in which case axioms $(E7)$, $(E8)$ and $(E9)$ would have involved this predicate symbol instead of A_1^2.

▷ For the rest of this chapter we shall be dealing with first order systems with equality in which A_1^2 stands for equality. The proof of Proposition 5.4 indicates how repetitive writing out proofs can become, and we can alleviate this somewhat by introducing the symbol $=$ into our language in place of A_1^2.

Notation. Write $t_1 = t_2$ in place of $A_1^2(t_1, t_2)$, where t_1 and t_2 are terms of $\mathscr{L}$.

Axioms $(E7)$, $(E8)$ and $(E9)$ can now be written in a simplified form, and in a way which makes their significance much clearer.

$(E7')$ $x_1 = x_1$

$(E8')$ $(t_k = u \rightarrow (f_i^n(t_1, \ldots, t_k, \ldots, t_n) = f_i^n(t_1, \ldots, u, \ldots, t_n)))$,
          $t_1, \ldots, t_n, u, f_i^n$ as in $(E8)$.

$(E9')$ $(t_k = u \rightarrow (A_i^n(t_1, \ldots, t_k, \ldots, t_n) \rightarrow A_i^n(t_1, \ldots, u, \ldots, t_n)))$,
          $t_1, \ldots, t_n, u, A_i^n$ as in $(E9)$.

The symbol $=$ is not the only one which we have introduced into the formal language in addition to the original alphabet of symbols. For example we use $(\exists x_i)$ as an abbreviation for $\sim(\forall x_i)\sim$, and we use $(\mathscr{A} \leftrightarrow \mathscr{B})$ as an abbreviation for $\sim((\mathscr{A} \rightarrow \mathscr{B}) \rightarrow \sim(\mathscr{B} \rightarrow \mathscr{A}))$. It is sometimes convenient to write $(\mathscr{A} \vee \mathscr{B})$ as an abbreviation for $(\sim\mathscr{A} \rightarrow \mathscr{B})$, and $(\mathscr{A} \wedge \mathscr{B})$ for $\sim(\mathscr{A} \rightarrow \sim\mathscr{B})$. These correspond with our intuitive ideas from Chapter 1, and the use of these new symbols clearly does not extend our formal systems in any way. It is a convenience to avoid lengthy repetition of symbols. In the different contexts which we are about to describe it is possible and sometimes desirable to extend the practice, and introduce further *defined symbols*. There is one useful one in particular which applies in any first order system with equality. This is the symbol for 'there exists a unique . . . such that'.

Notation. $(\exists_1 x_i)\mathscr{A}(x_i)$ is an abbreviation for the formula

$$(\exists x_i)(\mathscr{A}(x_i) \wedge (\forall x_j)(\mathscr{A}(x_j) \rightarrow x_i = x_j)).$$

Exercises

1   In Example 5.5, prove fully that every instance of axiom scheme $(E8)$ is true in the interpretation I.

2   Let S be a first order system with equality, and suppose that the closed wf. $\mathscr{A}$ is true in all normal models of S. Prove that $\mathscr{A}$ is true in all models of S.

3   Let $\mathscr{A}(x_1)$ be a *wf.* of $\mathscr{L}$ in which x_1 occurs free, let x_2 be free for x_1 in $\mathscr{A}(x_1)$, and let $\mathscr{A}(x_2)$ be the result of substituting x_2 for *one* of the free occurrences of x_1 in $\mathscr{A}(x_1)$. (N.B. this differs from our usual notational practice.) Prove that the *wf.*

$$(x_1 = x_2 \rightarrow (\mathscr{A}(x_1) \rightarrow \mathscr{A}(x_2)))$$

is a theorem of every extension of K in which $(E7')$, $(E8')$ and $(E9')$ are axioms. Deduce that the same result holds if $\mathscr{A}(x_2)$ is obtained by substituting x_2 for more than one free occurrence of x_1.

4   In the proof of Proposition 5.6, prove that the definition of $\hat{A}_i^n$ is well defined, i.e. that if $y_1, \ldots, y_n \in D_M$, $z_1, \ldots, z_n \in D_M$, and $[y_i] = [z_i]$ $(1 \le i \le n)$, then $\bar{A}_i^n(y_1, \ldots, y_n)$ holds if and only if $\bar{A}_i^n(z_1, \ldots, z_n)$ holds.

5   Describe how 'there exist precisely two $\ldots$ such that $\ldots$' can be expressed in a *wf.* of a first order system with equality.

6   Let $\mathscr{A}$ be a *wf.* of $\mathscr{L}$ involving the terms $t_1, \ldots, t_n$ 'free' in the sense that every occurrence of a variable in any of these terms is free in $\mathscr{A}$. Let S be an extension of $K_{\mathscr{L}}$ which includes the axioms $(E7')$, $(E8')$ and $(E9')$. Prove that, for any term u which contains no variable occurring bound in $\mathscr{A}$,

$$\vdash_S ((t_k = u) \rightarrow (\mathscr{A}(t_1, \ldots, t_k, \ldots, t_n) \rightarrow \mathscr{A}(t_1, \ldots, u, \ldots, t_n))).$$

5.3 The theory of groups

Group theory is perhaps the most familiar branch of mathematics which is based explicitly on a simple set of axioms, so let us use this 'mathematical context' to illustrate how mathematical systems arise as extensions of $K_{\mathscr{L}}$.

First we must describe an appropriate first order language $\mathscr{L}$, so let $\mathscr{L}_G$ be the first order language with the following alphabet of symbols:

variables $x_1, x_2, \ldots$
individual constant a_1 (identity)
function symbols f_1^1, f_1^2 (inverse, product)
predicate symbol $=$
punctuation $(,), ,$
logical symbols $\forall$, $\sim$, $\rightarrow$.

Now define $\mathscr{G}$ to be the extension of $K_{\mathscr{L}_G}$ whose proper axioms are $(E7)$, all appropriate instances of $(E8)$ and $(E9)$, and the following:

$(G1)$ $f_1^2(f_1^2(x_1, x_2), x_3) = f_1^2(x_1, f_1^2(x_2, x_3))$ (associative law)
$(G2)$ $f_1^2(a_1, x_1) = x_1$ (left identity)
$(G3)$ $f_1^2(f_1^1(x_1), x_1) = a_1$ (left inverse).

As previously, it does not matter whether universal quantifiers are included in these axioms for each free variable or not. An equivalent set of axioms would be the universal closures of these.

$(G1)$, $(G2)$ and $(G3)$ are merely translations of the usual group axioms. Normally $(G2)$ and $(G3)$ are stated in the form 'There exists a left identity' and 'For each element there exists a left inverse'. Our axioms here do not explicitly assert existence. They merely state that a_1 and $f_1^1(x_1)$, when interpreted in a model, must have the appropriate properties. To assert existence is unnecessary since in any model of this system there will be interpretations of a_1 and of f_1^1, and so the identity and inverse will automatically exist. Similarly, the group axiom concerning closure under the group operation is unnecessary here because the interpretation of f_1^2 in a model is necessarily a two place function with values in the domain of the model.

Given such a system of group theory, we can convert any standard proof from an algebra textbook of a result about elements of groups into a formal proof in the system. Such a procedure would have little practical use, for a formal proof in $\mathscr{G}$ is necessarily rather complicated, and the large numbers of purely manipulative steps would obscure the intuitive ideas involved, as the following example shows.

Example 5.8

In any group G with identity element e, $e(ee) = e$. Corresponding to this, let us give a formal proof in the system $\mathscr{G}$ of the wf.

$$f_1^2(a_1, f_1^2(a_1, a_1)) = a_1.$$

(1) $f_1^2(a_1, x_1) = x_1$ $(G2)$
(2) $(\forall x_1)(f_1^2(a_1, x_1) = x_1)$ (1), Generalisation
(3) $(\forall x_1)(f_1^2(a_1, x_1) = x_1) \to (f_1^2(a_1, a_1) = a_1)$ $(K5)$
(4) $f_1^2(a_1, a_1) = a_1$ (2), (3), MP
(5) $(\forall x_1)(f_1^2(a_1, x_1) = x_1) \to (f_1^2(a_1, f_1^2(a_1, a_1)) = f_1^2(a_1, a_1))$ $(K5)$
(6) $f_1^2(a_1, f_1^2(a_1, a_1)) = f_1^2(a_1, a_1)$ (2), (5), MP
(7) $(f_1^2(a_1, a_1) = a_1) \to (f_1^2(a_1, f_1^2(a_1, a_1)) = f_1^2(a_1, a_1) \to f_1^2(a_1, f_1^2(a_1, a_1)) = a_1)$ $(E9)$
(8) $(f_1^2(a_1, f_1^2(a_1, a_1)) = f_1^2(a_1, a_1) \to f_1^2(a_1, f_1^2(a_1, a_1)) = a_1)$ (4), (7), MP
(9) $f_1^2(a_1, f_1^2(a_1, a_1)) = a_1$ (6), (8), MP

In comparison with this, a standard proof of $e(ee) = e$ for any group is a triviality. More complicated results about groups are reflected in still more complicated formal proofs in $\mathscr{G}$. Particular examples are not very rewarding, but a further idea of the complications involved will be obtained by attempting to prove in $\mathscr{G}$ the wf.

$$f_1^2(x_1, a_1) = x_1,$$

corresponding to the property of groups that the left identity is also a right identity.

▷ It should be clear that any group G is a model of the system $\mathscr{G}$ provided that a_1 is interpreted as the identity element of G, f_1^1 as the inverse, f_1^2 as the group operation and = as equals. However, there are other models, as we shall see.

Example 5.9

Construct an interpretation I of the system $\mathscr{G}$ as follows. Let D_I be the set $\mathbb{Z}$ of integers, let a_1 be interpreted as 0, let

$$\bar{f}_1^1(x) = -x \quad \text{for } x \in \mathbb{Z},$$

and

$$\bar{f}_1^2(x, y) = x + y \quad \text{for } x, y \in \mathbb{Z},$$

and let = be interpreted by congruence (mod m), where m is some fixed positive integer. (Although we are using = as a symbol of $\mathscr{L}_G$, as we have seen above, it need not be interpreted always as actual equality.) I is a model of $\mathscr{G}$. To verify this we must show that every axiom of $\mathscr{G}$ is true in I. That $(K1)$–$(K6)$ are true requires no verification since they are logically valid. That $(E7)$, $(E8)$ and $(E9)$ are true is verified just as in Example 5.5. Let us look more closely at $(G1)$, $(G2)$ and $(G3)$.

$(G1)$ is interpreted as

$$(x + y) + z \equiv x + (y + z) \pmod{m}.$$

$(G2)$ is interpreted as

$$0 + x \equiv x \pmod{m}.$$

$(G3)$ is interpreted as

$$-x + x \equiv 0 \pmod{m}.$$

All of these are true statements, for any $x, y, z \in \mathbb{Z}$. Thus I is a model of $\mathscr{G}$. However, I is not a group. Indeed it involves the extraneous relation of congruence. However, the reader with some experience of group theory or number theory will realise that there is a group in the background waiting to be discovered. From the model I we can construct a normal model I^* by the procedure of Proposition 5.6. The domain of I^* is the set of congruence classes of integers (mod m), a_1 is interpreted by 0_m (the class containing 0), f_1^2 is interpreted by + (which is well-defined on congruence classes), f_1^1 is interpreted by 'additive inverse' (which is again well-defined), and = is interpreted by equals. I^* is a normal model, and it is a group.

▷ In general, any group is a normal model for the formal system of group theory, and conversely any normal model of the system is a group. So to make mathematical sense of the system we must restrict our attention to normal models. It is unfortunate, perhaps, but it is impossible to give axioms for equality which force the interpretation to be actual equality. It will always be possible to construct a model in which = is interpreted by some other equivalence relation.

The reason for constructing this formal system of group theory is not to provide any shortcuts or new methods for obtaining results about groups and their elements. As we have seen, the methods of proof within 𝒢 are so unwieldy as to be useless for this purpose. What we have gained by describing the system 𝒢 is that we have made precise and explicit all the assumptions and procedures which mathematicians use in the context of group theory, including the logical ones as well as the mathematical ones. In this way we have clarified this part of mathematics.

Groups have been treated in detail, and similar treatment can be given for other sorts of abstract algebraic systems, for example rings, fields, vector spaces, lattices, boolean algebras, etc. Each of these is known to be characterised by a finite set of axioms, and these can be easily translated into an appropriate formal language. Indeed every area of mathematics which is characterised by a set of axioms may be treated in a similar way. For example, Euclidean geometry can be based on a rather lengthy and complex set of axioms, and a formal system would have to include predicate letters intended to be interpreted by 'is a point', 'is a line', 'intersect' etc. Also an axiomatic system for the real numbers can be described, by means of the axioms for a complete ordered field.

There are two areas of mathematics which are particularly important when treated in this way. They are arithmetic and set theory. Each would require a whole book for a full treatment, but we shall merely try to explain why they have a special position. It is only within the framework of an explicit formal system that questions of consistency or of the relationships between different assumptions or of the position and use of fundamental assumptions can be clarified. Set theory serves as a foundation for all of mathematics, so its logical base is of over-riding importance. Arithmetic is an elementary fragment of mathematics, and its significance lies in the methods used to show that the search for a formal system which would enable any mathematical proposition to be tested must be fruitless. Any mathematical system in which ordinary arithmetic can be performed cannot be such a universal system, for the set of theorems of any consistent extension of arithmetic (in a sense which will be made precise) omits at least one true proposition. Some systems which are not extensions of arithmetic (e.g. the theory of groups) do not have this property. However a system which includes mathematical analysis or is intended to embrace mathematics as a whole will certainly include arithmetic, and so will suffer from this shortcoming. These matters will be discussed in some detail in Chapter 6.

Exercises

7 Describe a first order system $\mathscr{G}'$ of group theory, using a formal language which contains no individual constants. Do the same using a formal language which includes the individual constant a_1, but no function symbols.

8 A semigroup is a set on which is defined a binary operation which is associative. Describe a first order system $\mathscr{S}$ of semigroup theory such that the system $\mathscr{G}'$ of Exercise 7 is an extension of $\mathscr{S}$.

9 What is the effect on models of the system $\mathscr{G}'$ (from Exercise 7) if the system is altered by including in the language a single individual constant a_1 (but no additional axioms)? What is the effect of including in the language the whole sequence $a_1, a_2, \ldots$ of individual constants?

10 Describe a first order system whose normal models are all *infinite* groups. Is it possible for a first order system to have as its normal models all *finite* groups?

11 Describe in detail a first order system of ring theory, i.e. list the alphabet of symbols for an appropriate first order language, and write down a set of axioms and axiom schemes. Describe a model of this system which is not a ring.

12 Let $\mathscr{F}$ be a first order system of field theory. Normal models of this system are fields, which may have characteristic zero or any prime p. Prove that if a closed *wf.* $\mathscr{A}$ of the language of $\mathscr{F}$ is true in all fields of characteristic zero then there is a positive integer n such that $\mathscr{A}$ is true in all fields of characteristic p, with $p > n$.

13 Let $\mathscr{F}$ be as described in Exercise 12, and let $\mathscr{A}$ be a *wf.* such that for all p greater than some n, there is a field of characteristic p in which $\mathscr{A}$ is true. Prove that there is a field of characteristic zero in which $\mathscr{A}$ is true.

5.4 First order arithmetic

We develop the ideas involved in the arithmetic interpretation N first introduced in Chapter 3. The language $\mathscr{L}_N$ we take to include variables $x_1, x_2, \ldots$, the individual constant a_1 (for 0), the function letters f_1^1, f_1^2, f_2^2 (successor, sum and product), and the predicate symbol $=$, as well as punctuation, connectives and quantifier. Let us denote by $\mathscr{N}$ the first order system which is the extension of $K_{\mathscr{L}_N}$ obtained by including as additional axioms $(E7)$, all appropriate instances of $(E8)$ and $(E9)$, and the following six axioms and one axiom scheme.

(N1) $(\forall x_1) \sim (f_1^1(x_1) = a_1)$.
(N2) $(\forall x_1)(\forall x_2)(f_1^1(x_1) = f_1^1(x_2) \rightarrow x_1 = x_2)$.
(N3) $(\forall x_1)(f_1^2(x_1, a_1) = x_1)$.
(N4) $(\forall x_1)(\forall x_2)(f_1^2(x_1, f_1^1(x_2)) = f_1^1(f_1^2(x_1, x_2)))$.
(N5) $(\forall x_1)(f_2^2(x_1, a_1) = a_1)$.
(N6) $(\forall x_1)(\forall x_2)(f_2^2(x_1, f_1^1(x_2)) = f_1^2(f_2^2(x_1, x_2), x_1))$.
(N7) $\mathscr{A}(a_1) \rightarrow ((\forall x_1)(\mathscr{A}(x_1) \rightarrow \mathscr{A}(f_1^1(x_1))) \rightarrow (\forall x_1)\mathscr{A}(x_1))$,

for each wf. $\mathscr{A}(x_1)$ of $\mathscr{L}_N$ in which x_1 occurs free.

Notation. As yet we cannot know whether, for example, f_1^2 must, in any normal model, be interpreted as addition (or a function with the same properties as the sum function), but it will make the system $\mathscr{N}$ much clearer and the axioms above much easier to understand if we modify $\mathscr{L}_N$ immediately by using the symbols $+$, $\times$ and $'$ instead of f_1^2, f_2^2 and f_1^1 respectively. To be explicit, we shall write

$$t_1 + t_2 \quad \text{for } f_1^2(t_1, t_2),$$

$$t_1 \times t_2 \quad \text{for } f_2^2(t_1, t_2),$$

and

$$t' \quad \text{for } f_1^1(t),$$

where t, t_1, t_2 are any terms. Also we shall use the symbol 0 rather than a_1. The dangers of doing this must be emphasised once again. Having done it, we must not assume that these new symbols are necessarily always interpreted by the functions or objects which they normally represent.

Using these symbols, the axioms (N1)–(N7) may be rewritten as follows

(N1*) $(\forall x_1) \sim (x_1' = 0)$.
(N2*) $(\forall x_1)(\forall x_2)(x_1' = x_2' \rightarrow x_1 = x_2)$.
(N3*) $(\forall x_1)(x_1 + 0 = x_1)$.
(N4*) $(\forall x_1)(\forall x_2)(x_1 + x_2' = (x_1 + x_2)')$.
(N5*) $(\forall x_1)(x_1 \times 0 = 0)$.
(N6*) $(\forall x_1)(\forall x_2)(x_1 \times x_2' = (x_1 \times x_2) + x_1)$.
(N7*) $\mathscr{A}(0) \rightarrow ((\forall x_1)(\mathscr{A}(x_1) \rightarrow \mathscr{A}(x_1')) \rightarrow (\forall x_1)\mathscr{A}(x_1))$,

for each wf. $\mathscr{A}(x_1)$ in which x_1 occurs free.

Remarks 5.10

(*a*) The reader who is familiar with Peano's Postulates will recognise (N1), (N2) and (N7). Peano's Postulates are a set of axioms for the system of natural numbers which were first made explicit well before formal systems were studied as such. They are:

1. 0 is a natural number.
2. For each natural number n, there is another natural number n'.
3. For no natural number n is n' equal to 0.
4. For any natural numbers m and n, if $m' = n'$ then $m = n$.
5. For any set A of natural numbers containing 0, if $n' \in A$ whenever $n \in A$, then A contains every natural number.

Note that the first two postulates do not correspond with any of the axioms for our system $\mathcal{N}$. We do not need them because we have included symbols in the language $\mathscr{L}_N$ (0 and $'$, or a_1 and f_1^1) which must have interpretations in any model, so that in any model, an element $\bar{a}_1$ exists, and for each x there must be an element $\bar{f}_1^1(x)$.

(b) The correspondence between (N7) and the fifth of Peano's Postulates is not exact. Both are versions of the Principle of Mathematical Induction. However, because in $\mathcal{N}$ we are restricted to the use of the first order language $\mathscr{L}_N$, the axiom (N7) cannot be as strong or inclusive as Peano's fifth postulate. The reason is that Peano's fifth postulate contains a second order quantifier 'for any set A of natural numbers', which cannot be expressed in our first order language. The best we can do is use the notion of axiom *scheme*, so that we effectively have a quantifier in 'for every *wf.* $\mathscr{A}(x_1)$ in which x_1 occurs free'. Note that such a *wf.* $\mathscr{A}(x_1)$ determines a set in any interpretation, namely the set of all elements v_1 of the domain of the interpretation which satisfy $\mathscr{A}(x_1)$. (More precisely, the set of all elements v_1 of the domain of the interpretation such that every valuation v for which $v(x_1) = v_1$ satisfies $\mathscr{A}(x_1)$.)

If we think in the context of a model of $\mathcal{N}$, therefore, each instance of the axiom scheme (N7) corresponds to the assertion of Peano's fifth postulate in regard to one particular set. However, there is still an essential difference. The instances of axiom scheme (N7) form a countable set of *wf*s. of $\mathscr{L}_N$. Peano's fifth postulate is a statement about all sets of natural numbers, and the collection of all these is uncountable. Thus (N7) is a much restricted form of the Induction Principle, since it refers only to that countable collection of subsets of the domain of a model which can be 'represented' in the manner described above by *wf*s. of $\mathscr{L}_N$.

(c) Peano's Postulates contain no mention of sums or products. These functions can be defined in terms of the successor function, using the induction principle, but it is convenient to include symbols for these in the formal language. Having done this, axioms (N3)–(N6) are necessary to ensure that in any model the interpretations of these symbols have the required properties.

▷ There is a fundamental difference mathematically between this situation and the situation with groups. The formal system of group theory allowed many different normal models, namely all groups. The system $\mathcal{N}$ of arithmetic is intended to have only one normal model, namely the set of natural numbers, since it is properties of natural numbers which we hope will appear as theorems in the system. Whereas the group theorist may be concerned with general results which hold in all groups, the number theorist is concerned with results about a particular set, the set of natural numbers. It is a natural question to ask, therefore, whether there are any normal models of the system $\mathcal{N}$ other than the set of natural numbers. Another question which arises naturally is whether the system is strong enough, in the sense of having as theorems all *wfs.* which we would like to be theorems, i.e. all *wfs.* which correspond to true statements about natural numbers. These two questions are not unconnected, as we shall see shortly.

Some readers may be familiar with the standard proof that Peano's Postulates determine the set of natural numbers uniquely. Let N and M be 'models' of Peano's Postulates. Then $0 \in N$ and $0 \in M$. Let A be the set of elements of N which are elements of M. Then $0 \in A$. Also if $n \in A$, then $n \in N$ and $n \in M$, so $n' \in N$ and $n' \in M$, so $n' \in A$. Thus, by Peano's fifth postulate, A consists of all natural numbers, i.e. $A = N$, and so $N \subseteq M$. Similarly, $M \subseteq N$, and so $M = N$. In this proof, the fifth postulate is used essentially, and as we have noted above, $(N7)$ does not correspond exactly to this postulate. Indeed the above proof cannot be translated into a proof in $\mathcal{N}$. So there is no hope here of obtaining a negative answer to our first question about $\mathcal{N}$.

Let us now address ourselves to the question: Is $\mathcal{N}$ complete? i.e. is $\mathcal{A}$ or $(\sim\mathcal{A})$ always a theorem of $\mathcal{N}$, for each closed *wf.* $\mathcal{A}$ of $\mathscr{L}_N$? The significance of this question may not be obvious at first sight, but it has a bearing on both questions above. If $\mathcal{N}$ were not complete, then it would not be a strong enough system in the above sense, for then there would be a closed *wf.* $\mathcal{A}$ such that neither $\mathcal{A}$ nor $(\sim\mathcal{A})$ were theorems of $\mathcal{N}$. Now in any interpretation a closed *wf.* is either true or false, so in the interpretation N either $\mathcal{A}$ is true or $\mathcal{A}$ is false, and in the latter case $(\sim\mathcal{A})$ is true. Now the interpretation of $\mathcal{A}$ in N is a statement about natural numbers, and, intuitively, either $\mathcal{A}$ or $(\sim\mathcal{A})$ will have an interpretation which is a true statement about natural numbers. But neither $\mathcal{A}$ nor $(\sim\mathcal{A})$ is a theorem of $\mathcal{N}$. Thus if $\mathcal{N}$ were not complete then there would be a true statement about numbers whose corresponding *wf.* in $\mathcal{N}$ was not a theorem of $\mathcal{N}$. It would be desirable, and was part of the original aim in constructing the system $\mathcal{N}$, that all the *wfs.* which are true in the model N should be theorems of $\mathcal{N}$. However, if $\mathcal{N}$ were not complete then this could not be so.

Also, if there were a *wf.* $\mathscr{A}$ such that neither $\mathscr{A}$ nor $(\sim\mathscr{A})$ were theorems of $\mathcal{N}$, then (provided that $\mathcal{N}$ itself is consistent) we could obtain, just as we did at the end of Chapter 4, two distinct consistent extensions of $\mathcal{N}$ by adding first $\mathscr{A}$ as a new axiom and second $(\sim\mathscr{A})$ as a new axiom. Each of these extensions will have a normal model (Proposition 5.6), and these models are certainly models of $\mathcal{N}$ which must be essentially different, since in one $\mathscr{A}$ is true and in the other $(\sim\mathscr{A})$ is true. Thus if $\mathcal{N}$ were not complete there would necessarily be a normal model of $\mathcal{N}$ other than the intended one.

That $\mathcal{N}$ is not complete was one of the major results first obtained by Gödel. In fact he proved a much stronger result with this as a special case. Chapter 6 is devoted to the ideas and methods involved in the proof and examines some consequences of it. Before that we consider the other important 'mathematical context' mentioned earlier, namely formal set theory.

Exercise

14 A formal system $\mathcal{N}'$ of arithmetic could be specified as follows. The language contains $f_1^1, f_1^2, f_2^2, A_1^1, a_0, a_1, a_2, \ldots$ in addition to punctuation, connectives and quantifier as usual. The axioms are as for $\mathcal{N}$, with the addition of $f_1^1(a_i) = a_{i+1}$, for each $i > 0$. It is intuitively clear that N is a normal model of this system, provided that a_k is interpreted by $k - 1$, for each positive integer k. (The interpretation of a_0 is immaterial.) Now consider the system obtained from $\mathcal{N}'$ by including as additional axioms all *wfs*. $\sim(a_0 = a_i)$, for $i > 0$. By consideration of models, prove that this new system is consistent and therefore has a normal model. How does such a model differ from N?

5.5 Formal set theory

The foundations of mathematics are nowadays laid in the theory of sets, and since the beginning of this century mathematicians have investigated the basic assumptions that have to be made about sets (i.e. axioms) and the ways in which all of mathematics can be built upon these assumptions. The advantage of developing a formal theory of sets lies in making the assumptions explicit and providing an opportunity to criticise them and to explore interdependences between them. We shall describe one system of formal set theory. There are others, but ours is one of the standard ones, and it is perhaps easiest to describe, in terms of concepts we have already discussed. The reader who is unfamiliar with the set theoretic foundations of mathematics may find the axioms themselves difficult, but they are included here for the sake of completeness and in order to give some idea of their nature. What comes later does not depend on them. We do not have space to do more than describe the system and to point out some of the ways in which set theory develops from it.

The system which we describe is called *ZF*. The name derives from Ernst Zermelo, who first formulated a collection of axioms for set theory in 1905, and Abraham Fraenkel, who modified them in 1920.

The first order language which is appropriate for *ZF* contains variables, punctuation, connectives and quantifier as usual, and the predicate symbols $=$ and A_2^2 (no function letters or individual constants). A_2^2 is intended to be interpreted as $\in$, the relation of membership. Indeed, with the same warning as was given in the case of $\mathscr{L}_N$, we shall consider $\in$ as a symbol of the language, standing for A_2^2, and write $t_1 \in t_2$ in place of $A_2^2(t_1, t_2)$, for any terms t_1 and t_2. Notice that the lack of individual constants and function letters means that the only terms are the variables, and the only atomic formulas are of the form $x_i = x_j$ or $x_i \in x_j$. This may seem excessively restricting, but the axioms which we introduce will ensure that the formal system genuinely reflects the full generality of intuitive set theory, and we shall be able to introduce defined symbols corresponding to the standard notions of set theory, such as the empty set, union, power set, etc.

ZF is defined to be the extension of $K_{\mathscr{L}}$ (where $\mathscr{L}$ is as described above) obtained by including as axioms $(E7)$, all appropriate instances of $(E9)$ $(E8$ has no non-trivial instances), and $(ZF1)$ to $(ZF8)$ listed below.

$(ZF1)$ $(x_1 = x_2 \leftrightarrow (\forall x_3)(x_3 \in x_1 \leftrightarrow x_3 \in x_2))$.

This is called the Axiom of Extensionality, and its intended meaning is that two sets are equal if and only if they have the same elements. Note that the left to right implication is given already by $(E9)$, but it makes the significance of this axiom clearer if we include both implications here.

$(ZF2)$ $(\exists x_1)(\forall x_2) \sim (x_2 \in x_1)$.

This is the Null Set Axiom, since it guarantees the existence, in the intended interpretation, of a set with no members. It is a consequence of $(ZF1)$ that in any normal model there will be only one such set. We can thus introduce into the language the symbol $\varnothing$, to act as an individual constant, the *wf.* $(\forall x_2) \sim (x_2 \in \varnothing)$ being then the form that $(ZF2)$ takes.

Notation. We introduce the symbol $\subseteq$ as an abbreviation as follows:

$(t_1 \subseteq t_2)$ stands for $(\forall x_1)(x_1 \in t_1 \rightarrow x_1 \in t_2)$:

where t_1 and t_2 are any terms.

$(ZF3)$ $(\forall x_1)(\forall x_2)(\exists x_3)(\forall x_4)(x_4 \in x_3 \leftrightarrow (x_4 = x_1 \lor x_4 = x_2))$.

This is the Axiom of Pairing. Given any sets x and y there is a set z whose members are x and y. Again this axiom asserts existence, and it is convenient to introduce the symbols { and } into the language in order to denote the object whose existence is being asserted. $\{x_1, x_2\}$ will be regarded as a term, and $(ZF3)$ then asserts $x_4 \in \{x_1, x_2\} \leftrightarrow (x_4 = x_1 \vee x_4 = x_2)$.

$(ZF4)$ $(\forall x_1)(\exists x_2)(\forall x_3)(x_3 \in x_2 \leftrightarrow (\exists x_4)(x_4 \in x_1 \wedge x_3 \in x_4))$.

This is the Axiom of Unions. Given any set x, there is a set y which has as its members all members of members of x.

Notation. We denote by $\bigcup x_1$ the object whose existence is asserted in $(ZF4)$. This acts as a term, so $\bigcup$ acts as a one-place function symbol. We can then introduce $\cup$ by:

$(t_1 \cup t_2)$ stands for $\bigcup\{t_1, t_2\}$.

$(ZF5)$ $(\forall x_1)(\exists x_2)(\forall x_3)(x_3 \in x_2 \leftrightarrow x_3 \subseteq x_1)$.

This is the Power Set Axiom. Given any set x there is a set y which has as its members all the subsets of x.

$(ZF6)$ $(\forall x_1)(\exists_1 x_2)\mathcal{A}(x_1, x_2) \rightarrow$

$(\forall x_3)(\exists x_4)(\forall x_5)(x_5 \in x_4 \leftrightarrow (\exists x_6)(x_6 \in x_3 \wedge \mathcal{A}(x_6, x_5)))$,

for every *wf.* $\mathcal{A}(x_1, x_2)$ in which x_1 and x_2 occur free (and in which, we may suppose without loss of generality, the quantifiers $(\forall x_5)$ and $(\forall x_6)$ do not appear).

This is the Axiom Scheme of Replacement. If the *wf.* $\mathcal{A}$ determines a function, then for any set x, there is a set y which has as its members all the images of members of x under this function.

$(ZF7)$ $(\exists x_1)(\varnothing \in x_1 \wedge (\forall x_2)(x_2 \in x_1 \rightarrow x_2 \cup \{x_2\} \in x_1))$.

(N.B. $\{x_2\}$ is an abbreviation of $\{x_2, x_2\}$, defined above.)

This is the Axiom of Infinity. It asserts the existence, in any model, of an infinite set. If it were not included amongst the axioms there would be no way of ensuring that the formal system had any relevance to intuitive set theory which includes infinite sets.

$(ZF8)$ $(\forall x_1)(\sim x_1 = \varnothing \rightarrow (\exists x_2)(x_2 \in x_1 \wedge \sim(\exists x_3)(x_3 \in x_2 \wedge x_3 \in x_1)))$.

This is the Axiom of Foundation. Every non-empty set x contains a member which is disjoint from x. This is a technical axiom which is included in order to avoid anti-intuitive anomalies such as the possibility of a set being a member of itself.

ZF is a formal system of set theory. The axioms are chosen so that the interpretations of the formal symbols in normal models will behave as sets do. Some of the axioms have a stronger basis in intuition than others, but these eight have stood the test of time as representing basic truths about sets.

ZF can be used as a basis for mathematical analysis in the following way. On the assumption that it is a consistent system, we know that a normal model exists. It can be shown that in any such model there are sets with all the usual properties of the number systems. The details of this are lengthy and cannot be covered here. For example, a model for the system $\mathcal{N}$ of arithmetic can be defined as a subset of a model of ZF in the following way. $\varnothing$ has an interpretation in the model of ZF, $\bar{\varnothing}$, say. Then $\{\bar{\varnothing}\}$ is a different element of the model (the set whose only member is $\bar{\varnothing}$), and $\{\bar{\varnothing}, \{\bar{\varnothing}\}\}$ is another (this set has two elements, $\bar{\varnothing}$ and $\{\bar{\varnothing}\}$). This is the beginning of an inductive process, generating a sequence of sets. The general rule is: for each x in the sequence, its successor is $x \cup \{x\}$. It can be verified easily that the $(k+1)$st member of this sequence has k elements, and it is possible to *define* the natural number k as this $(k+1)$st member. We have already seen that the other arithmetic operations can be defined in terms of the successor function. The axioms $(N1) \ldots (N7)$ are then consequences of the definitions and the ZF axioms. Note that $(ZF7)$ is needed to ensure that the collection of all members of this sequence is an element of our normal model of ZF. In this way every normal model of ZF contains a normal model of $\mathcal{N}$.

The reader with a mathematical background may be familiar with the way in which the number systems of integers, rationals, reals and complex numbers may be constructed, starting from the natural members, by algebraic procedures. All of these procedures can be carried out within ZF. There is a lot of detailed verification required but the end result is confirmation that every normal model of ZF contains a set which looks and behaves like the set of complex numbers. (This set of course has a subset which looks and behaves like the set of real numbers.)

Besides the foundation of analysis on an axiomatic base, there was another stimulus at the turn of the century for the study of axiomatic set theory and this was in the intuitive justification (if any) for the use of certain principles in mathematics. Attention was then focussed on two particular principles, the axiom of choice (which was known to have several equivalent formulations) and the continuum hypothesis. It is quite illuminating to investigate some of the history of these principles since that time. Some mathematicians have regarded them as additional axioms of set theory, and others have regarded them either as suspect intuitively or even as falsehoods.

The axiom of choice is:

(AC)   For any non-empty set x there is a set y which has precisely one element in common with each member of x.

(Two of the best known equivalent formulations are: *Zorn's Lemma*; if each chain in a partially ordered set has an upper bound, then there is a maximal element of the set, and *The Well-Ordering Principle*; each set can be well-ordered.)

The continuum hypothesis is:

(CH)   Each infinite set of real numbers either is countable or has the same cardinal number as the set of all real numbers. (Two sets have the same cardinal number if there is a bijection between them.)

Because mathematicians were not in agreement about the acceptability of these two principles, the question of course was asked: are they true? The next question is: if these principles are to be demonstrated, on what principles ought such demonstrations to be based? Zermelo and Fraenkel (and others) listed what they thought to be the fundamental principles of set theory, and the question became: Can (AC) and (CH) be deduced as theorems of the system ZF of set theory, and if they can not, would it be consistent to include one or both as additional axioms?

Gödel (in 1938) answered one of these questions by technical consideration of the formal system of set theory. (AC) and (CH) are consistent with ZF. In other words, they may be included as additional axioms without introducing any contradiction. The idea is very simple. Under the assumption that ZF is consistent, he produced models in which (AC) and (CH) are true. Thus, by Propositions 4.43 and 4.44, the systems obtained by adding either (AC) or (CH) as additional axioms are both consistent. Incidentally he also showed that the system obtained by adding both (AC) and (CH) is consistent.

Much later Cohen (in 1963) answered the other question by proving that (AC) and (CH) can not be deduced as theorems of ZF. Again the idea is simple though the proof itself is not. Cohen produced models of ZF in which the negations of (AC) and (CH) are true. Now if (AC) and (CH) were theorems of ZF then they would be true in every model, but a wf. and its negation cannot both be true in any model.

The conclusion to all of this is that neither (AC) nor $\sim$(AC) is a theorem of ZF, and it would be consistent to include either of these as a new axiom. Likewise for (CH) and $\sim$(CH). The formal set theory has cleared the ground, and acceptance or non-acceptance of (AC) and (CH)

must be a matter for the intuition or for some yet undiscovered mathematical principle which may be accepted as a new axiom in future and which may confirm or deny (AC) and (CH). (Incidentally, Gödel's and Cohen's work also shows that (AC) and (CH) are independent of one another. Neither is a theorem of the system obtained from ZF by including the other as an additional axiom.)

The study of models of ZF, of different possible interpretations of $\in$, of the independence of the axioms, and of the relationship between the class of theorems of ZF and the class of 'true' statements in set theory is a substantial field of mathematics which we shall not enter, except to make the following remarks.

Exercise

15 How would the axioms for the formal system of set theory have to be modified if the language included the symbols a_1 (for $\varnothing$), f_1^2 (for $\{ , \}$), and A_3^2 (for $\subseteq$)?

Let the set of natural numbers be defined, as in the text, by $0 = \varnothing$, and $n + 1 = n \cup \{n\}$, within a model of ZF. Show that in the interpretation of the language of ZF whose domain is this set and in which $\in$ and $=$ are interpreted as actual set membership and actual equality respectively, the axioms $(ZF1)$, $(ZF2)$, $(ZF4)$ and $(ZF8)$ are true, and $(ZF3)$, $(ZF5)$ and $(ZF7)$ are false. Is $(ZF6)$ true or false?

5.6 Consistency and models

Any first order system is consistent if and only if it has a model. It is possible to argue, then, that the mathematical systems we have described are consistent because in each case we have been mirroring, in the axioms, the properties of an intended model. However, the perceptive reader may already have been concerned about a possible circularity in our arguments, which can be exemplified by the definition in Chapter 3 of an interpretation as a *set* with certain operations and relations. How can we talk of interpretations or models of ZF, the formal system of set theory, then, without a circularity? The answer is in the ideas previously mentioned of a metatheory embodying the assumptions which have to be made in order to prove results about formal systems. When we deal with the system $\mathcal{N}$, for example, it is possible to use, say, ZF as a metatheory since ZF 'contains' $\mathcal{N}$ in a sense already referred to. However, when we discuss ZF we have, so to speak, reached the end of the line. By its nature, ZF is to be appropriate to set theory and hence for all of mathematics. However, in order to study ZF we require mathematical methods which are not part of ZF. The notion of an interpretation of ZF can be defined only in terms of some intuitive metatheory concerning 'real' sets. The elements of a model of *ZF* are to be thought of as sets interpreting the symbols of *ZF*. However, the domain of a model of *ZF*, though it may be a 'real' set, cannot be a set in the same sense that the elements of that domain are, for it cannot be the interpretation of a symbol of *ZF*.

There are certainly intuitive and semantic difficulties in these matters, and it is because of this that demonstrations of consistency by means of models are generally held to be inadequate. The more respectable approach is the following. Given two first order systems S and S^*, we may attempt to show, on the assumption that a model exists for S^*, that a model for S can be constructed. This would give a proof of *relative consistency*. There is one situation where this is almost trivial.

Proposition 5.11

Let S^* be an extension of S ('extension' used as in Definition 4.34). Then if S^* is consistent, so is S.

Proof. Suppose that S^* is consistent, but S is not. Then $\vdash_S \mathscr{A}$ and $\vdash_S (\sim \mathscr{A})$, for some *wf.* $\mathscr{A}$ of S. But $\mathscr{A}$ is a *wf.* of S^* also, and any proof in S is also a proof in S^*, so $\vdash_{S^*} \mathscr{A}$ and $\vdash_{S^*} (\sim \mathscr{A})$, contradicting consistency of S^*.

$\triangleright$ This is the easiest situation to deal with. In cases where S^* is not an extension of S in this sense, for example where the languages of the two systems are different, the proof of relative consistency would be more difficult and may involve actual construction of a model of S from an assumed model of S^*. We have given one such construction, albeit sketchily, in showing that consistency of *ZF* implies consistency of $\mathcal{N}$.

It is not known whether *ZF* is consistent. Most logicians believe that it is, but any attempt to prove that it is consistent will lead to difficulties of the kind described above. Essentially, it would require the asumption of consistency of a system even more all-embracing than *ZF*. Certainly there would be no corresponding difficulties in the way of an attempt to disprove consistency. All that would be required for that would be an example of a *wf.* $\mathscr{A}$ such that both $\mathscr{A}$ and $(\sim \mathscr{A})$ are theorems of *ZF*. It is implicit in the above that no such *wf.* has yet been found. Seventy years of fruitless search is evidence that no such *wf.* exists, but it is in no way conclusive.

Finally let us note a result about models of *ZF* which is a consequence of Proposition 4.47, and which at first sight seems contradictory. *ZF* is a first order system. Under the assumption that *ZF* is consistent, Proposition 4.47 says that *ZF* has a *countable* model. Now uncountable sets exist, intuitively, so we would expect models of *ZF* to be uncountable, in order to contain such sets. This apparent paradox is called Skolem's Paradox, but we can escape from a direct contradiction by careful consideration of what constitutes a model, in the following way.

To be specific, axiom $(ZF5)$ is interpreted as 'given any set x, there is a set consisting of all subsets of x'. If x is an infinite countable set, then, according to the rules of set theory, x has uncountably many subsets. How can the set of all subsets of such a set x belong to a countable model? A countable model of ZF consists of sets. For any 'real' set x which belongs to the model (clearly there must be 'real' sets which do not belong to the model), axiom $(ZF5)$ asserts that all of the subsets of x which belong to the model constitute a set y which must also belong to the model. This set y will be countable, when regarded as a 'real' set, but it will be uncountable when regarded as an element of the model. An infinite set is uncountable if there is no bijection between it and the set of natural numbers. In the model there will be no bijection between y and the set of natural numbers (all 'real' such bijections will be missing from the model in the same way that some subsets of x are missing).

6
The Gödel Incompleteness Theorem

6.1 Introduction

Some indication has been given in Chapter 5 of the significance of the question: is the system $\mathcal{N}$ of formal arithmetic complete? In this chapter we shall describe Gödel's proof that $\mathcal{N}$ is not complete. The proof is highly technical, and indeed we shall omit some of the most technical parts, but there are two reasons for going into it at some length. The first is that this is a result which is fundamental to the study of formal systems and their value in the foundations of mathematics, and it is therefore interesting to see what sort of proof is required and to understand the concepts behind it. The second is that the proof introduces several new notions which have significance and usefulness beyond their immediate use here, and one of the purposes of this book is to introduce these notions. The most important of these are recursive functions and relations, Gödel numbering, and expressibility, and these are treated in the ensuing sections of this chapter for their own sake as well as with a view to their application in the proof of Gödel's theorem. The ideas of recursiveness are taken further in Chapter 7; the other sections of this chapter are not prerequisites for that later work. Because this chapter is structured in this way, the overall picture of the proof may be difficult to grasp from the later sections, and we shall start with a brief outline of the way the proof is built. The reader who is not particularly interested in the technicalities may omit the detail in Sections 2, 4 and 5 without prejudice to what comes after.

Gödel demonstrated the existence of a closed wf. $\mathcal{U}$ of $\mathcal{N}$ such that neither $\mathcal{U}$ nor $\sim\mathcal{U}$ is a theorem of $\mathcal{N}$. He did this by means of an explicit description of $\mathcal{U}$ and a demonstration that contradictions follow from the assumptions that $\mathcal{U}$ or $\sim\mathcal{U}$ are theorems of $\mathcal{N}$. The first technical idea is that of coding. A method is given (Gödel numbering) whereby each wf. and each sequence of wfs. is assigned a code number (a positive integer) such that the wf. or sequence of wfs. is easily recoverable from its code number. By means of this coding, statements about positive integers can be regarded as statements about code numbers of expressions in $\mathcal{N}$ or even about expressions in $\mathcal{N}$. For example we shall consider a two place relation on D_N which we denote by Pf, defined thus: $Pf(m, n)$ holds if and only if m is the code number of a proof in $\mathcal{N}$ of the wf. whose code number is n.

Now wfs. of $\mathcal{N}$ containing free variables are interpreted in N as relations among non-negative integers, so it is reasonable to ask: is there a wf. of $\mathcal{N}$ which is interpreted by the relation Pf? For technical reasons we ask a slightly different question. We first observe that in the language of $\mathcal{N}$ there exist terms $0, 0', 0'', \ldots$ standing for (i.e. interpreted in N by) the non-negative integers $0, 1, 2, \ldots$ These are denoted by $0, 0^{(1)}, 0^{(2)}, \ldots$ We then ask: is there a wf. $\mathcal{P}(x_1, x_2)$ of $\mathcal{N}$ with two free variables such that for every $m, n \in D_N$, either $\mathcal{P}(0^{(m)}, 0^{(n)})$ or $\sim\mathcal{P}(0^{(m)}, 0^{(n)})$ is a theorem of $\mathcal{N}$, depending on whether $Pf(m, n)$ holds in N or not. If there is, then Pf is said to be expressible in $\mathcal{N}$ by the wf. $\mathcal{P}$.

The next stage is to obtain a sufficient condition for relations to be expressible in $\mathcal{N}$, and this is where recursiveness comes in. For the definition see Section 6.3; we merely state here that a relation is expressible in $\mathcal{N}$ if it is recursive. It is easier to test for recursiveness than for expressibility, and it is through recursiveness that Gödel's proof shows that a particular relation is expressible in $\mathcal{N}$. The relation Pf above is in fact recursive, and this fact is used in the demonstration that the relation W is recursive also, where W is defined thus: $W(m, n)$ holds if and only if m is the code number of a wf. $\mathcal{A}(x_1)$ in which x_1 occurs free, and n is the code number of a proof in $\mathcal{N}$ of the wf. $\mathcal{A}(0^{(m)})$. It is worth a little effort to grasp the meaning of W, for it introduces a sort of 'self-reference' which is legitimate and is crucial to the proof.

This relation W is recursive and therefore is expressible in $\mathcal{N}$ by a wf. $\mathcal{W}$. From this $\mathcal{W}$, the wf. $\mathcal{U}$ is constructed by quite a simple procedure, and $\mathcal{U}$ has as its interpretation in N: for every $n \in D_N$, n is not the code number of a proof in $\mathcal{N}$ of the wf. $\mathcal{U}$. Thus $\mathcal{U}$ embodies a more startling form of self-reference, in that it can be regarded as asserting its own unprovability. It is by a careful distinction between wfs. and their code numbers and between the terms $0^{(n)}$ and the numbers they stand for that difficulties can be avoided. It is quite straightforward to show that if $\mathcal{U}$ were a theorem of $\mathcal{N}$ then a contradiction would follow, and it is only slightly more difficult to deduce a contradiction from the assumption that $\sim\mathcal{U}$ is a theorem of $\mathcal{N}$. Of course there is an unstated assumption that $\mathcal{N}$ is consistent (for otherwise both $\mathcal{U}$ and $\sim\mathcal{U}$ would automatically be theorems of $\mathcal{N}$), and in fact to show that $\sim\mathcal{U}$ is a theorem requires an assumption about $\mathcal{N}$ slightly stronger than consistency.

This, then, is the proof in outline. Some flesh will be put on the bones in the remainder of the chapter, and some extensions and consequences of the result are given in Section 6.5. Some proofs are omitted, and the reader is referred to the book by Mendelson for more of the technical details.

6.2 Expressibility

We continue here the study of the system $\mathcal{N}$ of arithmetic which has been described in Chapter 5, and of its intended model, N. One of the most important concepts originating in the study of logic, that of recursiveness, arises directly from the relationship between this formal system and the model N. The applications of recursive functions have been developed extensively in the past forty years, but originally they arose naturally from questions about expressibility, and this is where we shall start.

We have been concerned in previous chapters with the symbols of formal systems, and the ways in which they can be interpreted. In the present situation, we shall reverse this process. Let us start with the model N, whose domain is the set of natural numbers, which we shall continue to denote by D_N. First observe that the number 0, the successor function, addition, multiplication and equality are all represented in the system $\mathcal{N}$ in an obvious way by the symbols of $\mathcal{N}$. But consider, for example, the number 5. 5 is an element of D_N, but 5 is *not* a symbol of $\mathcal{N}$. However, 5 is the interpretation in N of a particular term, namely $0''''$ i.e. $f_1^1(f_1^1(f_1^1(f_1^1(f_1^1(a_1)))))$. Each natural number is the interpretation of a term of $\mathcal{N}$ in a similar way. We shall have occasion to use these terms, and so it is useful to introduce a more convenient notation than either of the above, which are both rather unwieldy.

Notation. $0^{(n)}$ is an abbreviation for 0 followed by n primes. Thus the *number* $n \in D_N$ is the interpretation in N of the *term* $0^{(n)}$. Trivially, $0^{(0)}$ stands for the individual constant 0 in $\mathcal{N}$.

It is important to note that, although we use $0^{(n)}$ to stand for a term of $\mathcal{N}$, the symbol n itself is *not* a symbol of $\mathcal{N}$ and the n which appears in $0^{(n)}$ *cannot* be substituted by a variable. We shall refer to the terms $0^{(n)}$ as *numeral terms*.

Proposition 6.1

Let $m, n \in D_N$.

(i)   If $m \neq n$, then $\vdash_{\mathcal{N}} \sim (0^{(m)} = 0^{(n)})$.

(ii)   If $m = n$, then $\vdash_{\mathcal{N}} (0^{(m)} = 0^{(n)})$.

Proof. (i) Suppose without loss of generality that $m > n$. Then there exists $k > 0$ such that $n = m + k$. Axiom $(N2^*)$ gives

$$\vdash_{\mathscr{N}} (0^{(m)} = 0^{(m+k)} \to 0^{(m-1)} = 0^{(m+k-1)}),$$

if $m > 0$, and repeated application, together with use of the rule HS, gives us

$$\vdash_{\mathscr{N}} (0^{(m)} = 0^{(m+k)} \to 0^{(0)} = 0^{(k)}).$$

(This is trivially the case if $m = 0$.)

Now $k > 0$, by hypothesis, so $k - 1 \in D_N$ and

$$\vdash_{\mathscr{N}} (0^{(k)} = (0^{(k-1)})').$$

(If this is not clear then consider it as

$$\vdash_{\mathscr{N}} (0^{\overbrace{'' \cdots '}^{k}} = (0^{\overbrace{'' \cdots '}^{k-1}})').)$$

Thus we have

$$\vdash_{\mathscr{N}} (0^{(m)} = 0^{(m+k)} \to 0^{(0)} = (0^{(k-1)})').$$

Using a tautology, we obtain

$$\vdash_{\mathscr{N}} (\sim(0^{(0)} = (0^{(k-1)})') \to \sim(0^{(m)} = 0^{(m+k)})).$$

But axiom $(N1^*)$ gives

$$\vdash_{\mathscr{N}} \sim(0^{(0)} = (0^{(k-1)})'),$$

and then MP yields

$$\vdash_{\mathscr{N}} \sim(0^{(m)} = 0^{(m+k)}) \text{ as required.}$$

(ii) Suppose that $m = n$. Then $0^{(m)}$ and $0^{(n)}$ are identical (they are the same term of $\mathscr{N}$). $(0^{(m)} = 0^{(n)})$ is thus an instance of axiom $(E7)$.

▷ We see that the set of terms of $\mathscr{N}$ contains a sequence $0, 0^{(1)}, 0^{(2)}, \ldots$ which is interpreted in N by the sequence of natural numbers. Now *wfs.* of $\mathscr{N}$ may involve these terms, and theorems of $\mathscr{N}$ involving these terms are interpreted in N as truths of arithmetic. The general question we shall investigate concerns the correspondence between truths of arithmetic and theorems of $\mathscr{N}$. Proposition 6.1 above points up a very limited part of this correspondence, but it leads us to the more general idea of expressibility. Another example should make this clearer.

Example 6.2

Consider the relation $\leq$ on the set D_N of natural numbers. Intuitively $m \leq n$ is the interpretation of the *wf.* $(\exists x_1)(0^{(m)} + x_1 = 0^{(n)})$, so $\leq$ is 'expressed' in $\mathcal{N}$ in this sense. But it is also expressed in a stronger sense, for

$$\text{if } m \leq n \text{ then } \underset{\mathcal{N}}{\vdash} (\exists x_1)(0^{(m)} + x_1 = 0^{(n)}),$$

and

$$\text{if } m \not\leq n \text{ then } \underset{\mathcal{N}}{\vdash} \sim (\exists x_1)(0^{(m)} + x_1 = 0^{(n)}).$$

Putting this in another way, the relation holds or does not hold of two natural numbers in D_N depending on whether a particular *wf.* or its negation is a theorem of the formal system $\mathcal{N}$. We shall omit the details of verification that all the required formulas are theorems of $\mathcal{N}$ in this particular example.

Definition 6.3

A k-place relation R on natural numbers is *expressible in* $\mathcal{N}$ if there is a *wf.* $\mathcal{A}(x_1, \ldots, x_k)$ with k free variables such that, for any $n_1, n_2, \ldots, n_k \in D_N$,

(i)   if $R(n_1, \ldots, n_k)$ holds in N then $\underset{\mathcal{N}}{\vdash} \mathcal{A}(0^{(n_1)}, \ldots, 0^{(n_k)})$,

and

(ii)   if $R(n_1, \ldots, n_k)$ does not hold in N

then $\underset{\mathcal{N}}{\vdash} \sim \mathcal{A}(0^{(n_1)}, \ldots, 0^{(n_k)})$.

Remark 6.4

(*a*) Proposition 6.1 says that the relation of equality in N is expressible in $\mathcal{N}$ in this precise sense.

(*b*) If $\mathcal{N}$ were known to be complete, Definition 6.3 could be more simply stated, combining parts (i) and (ii) into a single 'if and only if' statement, for we would know that either $\mathcal{A}(0^{(n_1)}, \ldots, 0^{(n_k)})$ or $\sim \mathcal{A}(0^{(n_1)}, \ldots, 0^{(n_k)})$ would necessarily be a theorem of $\mathcal{N}$. However, it is possible that for some *wf.* $\mathcal{A}$ and numbers $n_1, \ldots, n_k$, neither $\mathcal{A}(0^{(n_1)}, \ldots, 0^{(n_k)})$ nor $\sim \mathcal{A}(0^{(n_1)}, \ldots, 0^{(n_k)})$ is a theorem of $\mathcal{N}$. Thus both conditions in Definition 6.3 are necessary.

(*c*) The argument in (*b*) above shows the possibility that not every *wf.* of $\mathcal{N}$ which involves free variables 'expresses' a relation in this way, whereas certainly every such *wf.* is interpreted as a relation in N.

(*d*) Sets of natural numbers can be regarded as unary relations in this context. If A is a subset of D_N, then '$\in A$' is a unary relation on D_N which may or may not be expressible in $\mathcal{N}$ in the above sense. For example, let $\mathscr{A}$ be the set of even numbers. Then '$\in A$' is the interpretation of the *wf*.

$$(\exists x_2)(x_2 \times 0^{(2)} = x_1).$$

The reader may consider if for each $m \in D_N$, either $(\exists x_2)(x_2 \times 0^{(2)} = 0^{(m)})$ or $\sim(\exists x_2)(x_2 \times 0^{(2)} = 0^{(m)})$ must be a theorem of $\mathscr{N}$, i.e. whether the relation 'is even' is expressible in $\mathscr{N}$.

(*e*) A function is a particular kind of relation. In general, a $(k+1)$-place relation R on D_N is a function if, for every $n_1, \ldots, n_k \in D_N$, there is precisely one $n_{k+1} \in D_N$ such that $R(n_1, \ldots, n_k, n_{k+1})$ holds. Now in considering whether a function (as a relation) is expressible in $\mathscr{N}$ it is relevant to consider whether this single valuedness property is shared by the *wf*. of $\mathscr{N}$ concerned.

Definition 6.5

A k-place function on D_N (i.e. a function $D_N^k \to D_N$) is *representable in $\mathscr{N}$* if it is expressible (as a $(k+1)$-place relation) in $\mathscr{N}$ by a *wf*. $\mathscr{A}$ with $k+1$ free variables, such that for every $n_1, \ldots, n_k \in D_N$,

$$\underset{\mathscr{N}}{\vdash} (\exists_1 x_{k+1}) \mathscr{A}(0^{(n_1)}, \ldots, 0^{(n_k)}, x_{k+1}).$$

Remark 6.6

Again we must remind ourselves of the distinction between the terms $0^{(n)}$ and the elements of D_N. At first sight it might appear that the condition

$$\underset{\mathscr{N}}{\vdash} (\forall x_1) \ldots (\forall x_k)(\exists_1 x_{k+1}) \mathscr{A}(x_1, \ldots, x_k, x_{k+1}) \qquad (*)$$

is equivalent to the condition given in the Definition. This is not so, and the easiest way to see why it need not be so is to consider a possible different model of $\mathscr{N}$. This model would have to contain interpretations for all the terms $0, 0^{(1)}, 0^{(2)}, \ldots$ but it would contain other elements as well. The interpretation of (*) would be an assertion about these other elements, and so would be a stronger assertion than the conjunction of the interpretations of $(\exists_1 x_{k+1}) \mathscr{A}(0^{(n_1)}, \ldots, 0^{(n_k)}, x_{k+1})$ for every $n_1, \ldots, n_k$.

Example 6.7

The function $f: D_N^2 \to D_N$ given by $f(m, n) = m + n$ is representable in $\mathscr{N}$.

Let $\mathscr{A}(x_1, x_2, x_3)$ be the *wf*. $x_3 = x_1 + x_2$. It is required to show the following, for each $m, n, p \in D_N$:

(i)   if $p = m + n$ then $\underset{\mathscr{N}}{\vdash} 0^{(p)} = 0^{(m)} + 0^{(n)}$,

(ii)   if $p \neq m + n$ then $\underset{\mathscr{N}}{\vdash} \sim(0^{(p)} = 0^{(m)} + 0^{(n)})$,

and

(iii) $\vdash_{\mathcal{N}} (\exists_1 x_3)(x_3 = 0^{(m)} + 0^{(n)})$.

We can demonstrate these as follows. Let $m, n \in D_N$. Then $\vdash_{\mathcal{N}} 0^{(m)} + 0^{(n)} = 0^{(m+n)}$. If $n = 0$ this is just axiom $(N3^*)$. If $n > 0$, write $0^{(n)}$ as $(0^{(n-1)})'$. By $(N4^*)$, then,

$$\vdash_{\mathcal{N}} (0^{(m)} + 0^{(n)}) = (0^{(m)} + 0^{(n-1)})'.$$

Repeat this process as often as necessary, to obtain

$$\vdash_{\mathcal{N}} (0^{(m)} + 0^{(n)}) = (0^{(m)} + 0)^{\overbrace{\prime \cdots \prime}^{n}}$$

i.e.

$$\vdash_{\mathcal{N}} (0^{(m)} + 0^{(n)}) = (0^{\overbrace{\prime \cdots \prime}^{m}})^{\overbrace{\prime \cdots \prime}^{n}}$$

i.e.

$$\vdash_{\mathcal{N}} (0^{(m)} + 0^{(n)} = 0^{(m+n)}).$$

(i) and (ii) are now immediate consequences of Proposition 6.1. For (iii), we must show that

$$\vdash_{\mathcal{N}} (\exists_1 x_3)(x_3 = 0^{(m)} + 0^{(n)} \wedge (\forall x_i)(x_i = 0^{(m)} + 0^{(n)} \rightarrow x_i = x_3)).$$

Now

$$\vdash_{\mathcal{N}} (0^{(m+n)} = 0^{(m)} + 0^{(n)}) \wedge (\forall x_i)(x_i = 0^{(m)} + 0^{(n)} \rightarrow x_i = 0^{(m+n)}).$$

(Details are omitted, and are not difficult.) Using the result of Exercise 3.21, we obtain the desired result.

Example 6.8

The function $f: D_N \rightarrow D_N$ given by $f(m) = 2m$ is representable in $\mathcal{N}$.

Let $\mathcal{A}(x_1, x_2)$ be the wf. $x_2 = x_1 \times 0^{(2)}$. We must prove the following for each $m, n \in D_N$;

(i)   if $n = 2m$ then $\vdash_{\mathcal{N}} 0^{(n)} = 0^{(m)} \times 0^{(2)}$,

(ii)   if $n \neq 2m$ then $\vdash_{\mathcal{N}} \sim(0^{(n)} = 0^{(m)} \times 0^{(2)})$,

and

(iii) $\vdash_{\mathcal{N}} (\exists_1 x_2)(x_2 = 0^{(m)} \times 0^{(2)})$.

For (i), suppose that $n = 2m$. We give an outline of a proof in $\mathcal{N}$ of $0^{(n)} = 0^{(m)} \times 0^{(2)}$.

$$0^{(m)} \times 0^{(2)} = 0^{(m)} \times 0'' \qquad\qquad \text{notation}$$
$$= (0^{(m)} \times 0') + 0^{(m)} \qquad\qquad (N6^*)$$
$$= ((0^{(m)} \times 0) + 0^{(m)}) + 0^{(m)} \qquad\qquad (N6^*)$$
$$= (0 + 0^{(m)}) + 0^{(m)} \qquad\qquad (N5^*)$$
$$= 0^{(m)} + 0^{(m)} \qquad\qquad (N3^*)$$
$$= 0^{(m+m)} \qquad\qquad \text{by previous example}$$
$$= 0^{(2m)}$$
$$= 0^{(n)}.$$

For (ii), suppose that $n \neq 2m$. Then $\underset{\mathcal{N}}{\vdash} \sim(0^{(2m)} = 0^{(n)})$ by Proposition 6.1, and $\underset{\mathcal{N}}{\vdash} (0^{(2m)} = 0^{(m)} \times 0^{(2)})$ as above. Using axiom $(E9')$, then, we obtain $\underset{\mathcal{N}}{\vdash} \sim(0^{(m)} \times 0^{(2)} = 0^{(n)})$.

For (iii), the demonstration is just as in the previous example.

Example 6.9

The function Z of two places, defined by $Z(m, n) = 0$ for every $m, n \in D_N$, is representable in $\mathcal{N}$.

Let $A(x_1, x_2, x_3)$ be the *wf*.

$$(x_1 = x_1 \wedge x_2 = x_2 \wedge x_3 = 0).$$

Again verification is required that certain *wfs*. are theorems of $\mathcal{N}$:

(i)   if $Z(m, n) = p$ then $\underset{\mathcal{N}}{\vdash} (0^{(m)} = 0^{(m)} \wedge 0^{(n)} = 0^{(n)} \wedge 0^{(p)} = 0)$,

(ii)   if $Z(m, n) \neq p$ then $\underset{\mathcal{N}}{\vdash} \sim(0^{(m)} = 0^{(m)} \wedge 0^{(n)} = 0^{(n)} \wedge 0^{(p)} = 0)$,

and

(iii) $\underset{\mathcal{N}}{\vdash} (\exists_1 x_3)(0^{(m)} = 0^{(m)} \wedge 0^{(n)} = 0^{(n)} \wedge x_3 = 0)$,

for every $m, n, p \in D_N$.

For (i), if $Z(m, n) = p$ then $p = 0$, and each of $0^{(m)} = 0^{(m)}$, $0^{(n)} = 0^{(n)}$, $0^{(p)} = 0$ is a theorem of $\mathcal{N}$.

For (ii), if $Z(m, n) \neq p$, then $p \neq 0$, and so $\sim(0^{(p)} = 0)$ is a theorem of $\mathcal{N}$ (axiom $(N1^*)$). It follows that

$$\underset{\mathcal{N}}{\vdash} \sim(0^{(m)} = 0^{(m)} \wedge 0^{(n)} = 0^{(n)} \wedge 0^{(p)} = 0).$$

To verify (iii), it is sufficient to show that $\underset{\mathcal{N}}{\vdash} (\exists_1 x_3)$ $(x_3 = 0)$, i.e. that $\underset{\mathcal{N}}{\vdash} (\exists x_3)(x_3 = 0 \wedge (\forall x_i)(x_i = 0 \rightarrow x_i = x_3))$. This holds because

$$\vdash_{\mathcal{N}} (0 = 0 \wedge (\forall x_i)(x_i = 0 \to x_i = 0)).$$

Again here we use the result of Exercise 3.21.

▷ The first general question one may ask about functions representable in $\mathcal{N}$ is whether there are functions which are not so representable. Given a function on D_N it may be hard to verify that it is representable, and possibly harder to verify that it is not. Some of the difficulties are exemplified in the above. General results will therefore be useful.

Proposition 6.10

Not all functions on D_N are representable in $\mathcal{N}$.

Proof. Compare the argument in Remark 5.10(*b*). The set of *wfs*. on $\mathcal{N}$ is countable, so the set of functions representable in $\mathcal{N}$ is countable. But there are uncountably many functions on D_N. Hence there are functions on D_N which are not representable in $\mathcal{N}$.

Corollary 6.11

Not all relations on D_N are expressible in $\mathcal{N}$.

Proof. Similar to the above.

▷ The next question it seems logical to ask is: can we characterise the set of functions (relations) on D_N which are representable (expressible) in $\mathcal{N}$. The answer is an important result and is one of the keys to Gödel's theorem about the incompleteness of the system $\mathcal{N}$.

Proposition 6.12

A function (relation) on D_N is representable (expressible) in $\mathcal{N}$ if and only if it is recursive.

This of course means nothing to us yet, and indeed its full proof is beyond our scope, but once we have defined and described recursive functions and relations we shall see how important it is.

Exercises

1 Prove that, for any $m, n \in D_N$, if $m \leq n$, then

$$\vdash_{\mathcal{N}} (\exists x_1)(0^{(m)} + x_1 = 0^{(n)}).$$

2 Prove that, for any $m, n \in D_N$, if $m > n$, then

$$\vdash_{\mathcal{N}} \sim(\exists x_1)(0^{(m)} + x_1 = 0^{(n)}).$$

3 Write down w/s. of the system $\mathcal{N}$ whose interpretations in N are the following.

(a) N is not a prime number;

(b) m and n have no common factor;

(c) $m = \min(p, q)$;

(d) every $m \in D_N$ has a prime factor;

(e) $sg(m) = n$ (for definition see p. 140).

4 Prove that the following functions on D_N are representable in $\mathcal{N}$, without using Proposition 6.12.

(a) sg (for definition see p. 140);

(b) f, where $f(n) = n + 3$;

(c) rm_2, where $rm_2(n)$ is the remainder on division of n by 2.

5 Show that a function $f: D_N \rightarrow D_N$ is representable in $\mathcal{N}$ if there is a wf. $\mathcal{A}$ of $\mathcal{N}$ such that

(i) If $f(m) = n$ then $\vdash_{\mathcal{N}} \mathcal{A}(0^{(m)}, 0^{(n)})$,

and

(ii) For each $m \in D_N$, $\vdash_{\mathcal{N}} (\exists_1 x_2) \mathcal{A}(0^{(m)}, x_2)$.

(This extends to functions of any number of variables.)

6 Prove that the projection function $p_i^k: D_N^k \rightarrow D_N$, given by $p_i^k(n_1, \ldots, n_k) = n_i$ $(n_1, \ldots, n_k \in D_N)$, is representable in $\mathcal{N}$, for each $i, k > 0$.

6.3 Recursive functions and relations

The class of recursive functions is defined (without any reference to the system $\mathcal{N}$) in the following way. Certain easily defined functions are recursive, and all functions obtained from these by application of three rules are also recursive.

The basic functions are:

1. The zero function $z: D_N \rightarrow D_N$, given by $z(n) = 0$ for every $n \in D_N$.

2. The successor function $s: D_N \rightarrow D_N$, given by $s(n) = n + 1$ for every $n \in D_N$.

3. The projection functions $p_i^k: D_N^k \rightarrow D_N$, given by $p_i^k(n_1, \ldots, n_k) = n_i$, for every $n_1, \ldots, n_k \in D_N$. N.B., p_1^1 is the identity function.

The three rules are:

I. Composition. If $g: D_N^j \rightarrow D_N$ and $h_i: D_N^k \rightarrow D_N$ for $1 \le i \le j$, then $f: D_N^k \rightarrow D_N$, defined by

$$f(n_1, \ldots, n_k) = g(h_1(n_1, \ldots, n_k), \ldots, h_j(n_1, \ldots, n_k)),$$

is obtained by composition from g and $h_1, \ldots, h_j$.

II. Recursion. If $g: D_N^k \rightarrow D_N$ and $h: D_N^{k+2} \rightarrow D_N$, then the function $f: D_N^{k+1} \rightarrow D_N$, defined by

$$f(n_1, \ldots, n_k, 0) = g(n_1, \ldots, n_k),$$

and

$$f(n_1, \ldots, n_k, n+1) = h(n_1, \ldots, n_k, n, f(n_1, \ldots, n_k, n)),$$

is said to be obtained by recursion from g and h. Note that here the $n_1, \ldots, n_k$ are parameters which do not affect the definition, and we shall allow these to be absent. The function f defined by

$$f(0) = a \quad \text{(a fixed member of } D_N\text{)},$$

and

$$f(n+1) = h(n, f(n)),$$

is likewise obtained by recursion.

III. Least Number Operator. Let $g: D_N^{k+1} \to D_N$ be any function with the property that for each $n_1, \ldots, n_k \in D_N$ there is at least one $n \in D_N$ such that $g(n_1, \ldots, n_k, n) = 0$. Then the function $f: D_N^k \to D_N$, defined by

$$f(n_1, \ldots, n_k) = \text{least number } n \in D_N$$
$$\text{such that } g(n_1, \ldots, n_k, n) = 0,$$

is said to be obtained from g by use of the least number operator.

Notation. The least number n such that $g(n_1, \ldots, n_k, n) = 0$ is denoted by $\mu n[g(n_1, \ldots, n_k, n) = 0]$.

Remark 6.13

The condition on g that for any $n_1, \ldots, n_k$ there is at least one n such that $g(n_1, \ldots, n_k, n) = 0$ is clearly necessary to ensure that the function f is a total function, i.e. has a value for each k-tuple of natural numbers. We shall later have occasion to allow use of the least number operator without this condition. This will obviously lead to the introduction of partial functions. For the moment, however, the condition stays, and all our functions are total functions.

Example 6.14

(a) The function $f: D_N^3 \to D_N$ given by $f(m, n) = m^2 + mn$ is obtained by composition from the addition, multiplication and projection functions thus (where f_1 denotes sum and f_2 denotes product):

$$f(m, n) = f_2(p_1^2(m, n), f_1(m, n)).$$

(b) The function $g: D_N^3 \to D_N$ given by $g(m, n, p) = n^2$ is obtained by composition thus:

$$g(m, n, p) = f_2(p_2^3(m, n, p), p_2^3(m, n, p)),$$

where f_2 denotes product.

(c) The addition function is obtained by recursion from p_1^1 and the composition of s with p_3^3 thus:

$$f_1(m, 0) = p_1^1(m)$$

$$f_1(m, n+1) = s(p_3^3(m, n, f_1(m, n))).$$

(d) Similarly the multiplication function is obtained by recursion from the addition function.

(e) Let $f(n)=$ least number q such that $n+q \equiv 0 \pmod{p}$ $(n, p, q \in D_N)$. Then f is obtained by the least number operator from the function g, where

$$g(n, p, q) = \text{the remainder when } n+q \text{ is divided by } p.$$

Definition 6.15

A function on D_N is *recursive* if it can be obtained from functions 1, 2, 3 above by a finite number of applications of the rules I, II, III. The class of recursive functions is thus the smallest class of functions on D_N which contains all of 1, 2, 3 and is closed under the application of rules I, II, III.

A function is *primitive recursive* if it can be obtained from functions 1, 2, 3 by a finite number of applications of rules I and II only. The primitive recursive functions form a strictly smaller class of functions than the recursive functions (this requires proof, which we shall not go into). They are important in certain branches of the subject, but we shall not have need to consider them specifically.

Example 6.16

(a) The sum function is (primitive) recursive. To see this, return to Example 6.14(c), where the sum function is obtained by recursion from a projection function and the successor function.

(b) The product function is (primitive) recursive. To see this define $f_2: D_N^2 \to D_N$ as follows:

$$f_2(m, 0) = z(m)$$

$$f_2(m, n+1) = h(m, n, f_2(m, n)),$$

where $h(m, n, p) = f_1(p_3^3(m, n, p), p_1^3(m, n, p))$. f_2 is here defined by recursion from z, which is recursive, and h, which is recursive, being obtained by composition from f_1, p_3^3, and p_1^3.

(c) Example 6.14(b) is of a recursive function.

(d) The function $f: D_N \to D_N$ given by $f(n) = n!$ is (primitive) recursive. For f is defined by

$$f(0) = 1$$

$$f(n+1) = f_2(s(n), f(n)),$$

and so is obtained by recursion (and composition) from the successor and multiplication functions, which are (primitive) recursive.

(e) All constant functions are recursive. The constant function with value k can be defined from the projection function p_2^2 thus:

$$f(0) = k$$

$$f(n+1) = p_2^2(n, f(n)).$$

Constant functions of more than one place are seen to be recursive by use of other projection functions.

(f) The functions sg, $\overline{sg}$: $D_N \to D_N$ given by

$$sg(n) = \begin{cases} 0 & \text{if } n = 0 \\ 1 & \text{if } n \neq 0 \end{cases} \qquad \overline{sg}(n) = \begin{cases} 1 & \text{if } n = 0 \\ 0 & \text{if } n \neq 0 \end{cases}$$

are recursive. For

$$sg(0) = 0$$

$$sg(n+1) = 1 \quad \text{(the constant function)},$$

and

$$\overline{sg}(0) = 1$$

$$\overline{sg}(n+1) = 0.$$

▷ It may appear that this definition is just as unwieldy to apply as the definition of function representable in $\mathcal{N}$, but this is not so, and its inductive nature enables one quite quickly and easily to build up a large collection of functions known to be recursive. However, there is another benefit which comes with discussion of recursiveness, and that concerns *computability*, of which we shall say more later.

The notion of recursiveness is extended to relations by means of the idea of characteristic function.

Definition 6.17

Let R be a k-place relation on D_N. The *characteristic function* of R (denoted by C_R) is defined by

$$C_R(n_1, \ldots, n_k) = \begin{cases} 0 & \text{if } R(n_1, \ldots, n_k) \quad \text{holds} \\ 1 & \text{if } R(n_1, \ldots, n_k) \quad \text{does not hold.} \end{cases}$$

Definition 6.18

A relation R on D_N is *recursive* if its characteristic function is a recursive function.

Example 6.19

The binary relation R, where $R(m, n)$ holds if and only if $m + n$ is even, is recursive.

To prove this, we must show that the function f defined by

$$f(m, n) = \begin{cases} 0 & \text{if } m + n \text{ is even} \\ 1 & \text{if } m + n \text{ is odd} \end{cases}$$

is a recursive function. Let $rm_2 : D_N \to D_N$ be the function given by $rm_2(n) = $ remainder on division of n by 2. rm_2 is recursive, for

$$rm_2(0) = 0$$

$$rm_2(n + 1) = \overline{sg}(rm_2(n)) = \overline{sg}(p_2^2(n, rm_2(n))).$$

Now $f(m, n) = rm_2(m + n)$, so f is recursive, by composition, since $+$ and rm_2 are recursive.

Example 6.20

The relation $\leq$ is recursive.

To prove this we must show that the function g defined by

$$g(m, n) = \begin{cases} 0 & \text{if } m \leq n \\ 1 & \text{if } m > n \end{cases}$$

is recursive. This requires a number of stages. First, the function p, defined by

$$p(n) = \begin{cases} n - 1 & \text{if } n > 0 \\ 0 & \text{if } n = 0, \end{cases}$$

is recursive. To see this,

$$\left. \begin{array}{l} p(0) = 0 \\ p(n + 1) = n \end{array} \right\} \text{ using recursion.}$$

Next, the function $\dotdiv$, defined by

$$m \dotdiv n = \begin{cases} m - n & \text{if } m \geq n \\ 0 & \text{if } n > m, \end{cases}$$

is recursive. To see this,

$$\left. \begin{array}{l} m \dotdiv 0 = m \\ m \dotdiv (n + 1) = p(m \dotdiv n) \end{array} \right\} \text{ using recursion.}$$

Note that the subtraction function cannot be discussed here because it leads to negative values, and our domain of numbers consists of just the non-negative integers. The function denoted by $\dotdiv$ is a modified subtraction.

Now we can see how to define g as required.

$g(m, n) = sg(m \div n)$, using composition.

▷ The definition of recursiveness may apply to sets of numbers as well. If $A \subseteq D_N$, we say that A is recursive if the characteristic function of A is recursive. (The characteristic function of a set A is just the characteristic function, as defined above, of the relation '$\in A$'.)

Example 6.21

(a) The set D_N is recursive, as its characteristic function is the zero function, which is recursive.

(b) $\varnothing$ is recursive, as its characteristic function is a constant function.

(c) The set of even numbers is recursive. To see this we must show that the function h given by

$$h(n) = \begin{cases} 0 & \text{if } n \text{ is even} \\ 1 & \text{if } n \text{ is odd} \end{cases}$$

is recursive. h is just the function rm_2 defined in Example 6.19.

Proposition 6.22

If R and S are recursive k-place relations, then the relations $\bar{R}$, $R \wedge S$ and $R \vee S$ are recursive. ($\bar{R}$ is the relation which holds for a given k-tuple if and only if R does not hold for that k-tuple. $R \wedge S$ holds of a given k-tuple if and only if R and S both hold. $R \vee S$ holds of a given k-tuple if and only if either R or S holds.)

Proof. The characteristic function of $\bar{R}$ is $\overline{sg}(C_R)$, so is recursive given that C_R is recursive. Also

$$C_{R \wedge S}(n_1, \ldots, n_k) = sg(C_R(n_1, \ldots, n_k) + C_S(n_1, \ldots, n_k))$$

and

$$C_{R \vee S}(n_1, \ldots, n_k) = C_R(n_1, \ldots, n_k) \times C_S(n_1, \ldots, n_k).$$

Therefore $C_{R \wedge S}$ and $C_{R \vee S}$ are recursive, given that C_R and C_S are recursive.

Corollary 6.23

For any recursive sets A and B, the complement of A, the intersection of A and B, and the union of A and B are recursive sets.

Proof. This result is just a special case of the proposition, for sets A and B have characteristic functions which are the characteristic functions of the relations '$\in A$' and '$\in B$'.

▷ One may proceed from here to demonstrate the recursiveness of particular functions, relations and sets, and indeed it is necessary to do so to prove Proposition 6.12 and the incompleteness of the system $\mathcal{N}$. In fact it is more difficult to find a function or relation which is *not* recursive, for virtually all easily describable functions and relations are recursive. We shall carry this process some way, in order to give a feel for the procedures involved, and some idea of the difficulties of describing a non-recursive function or relation will emerge.

Proposition 6.24

Every singleton subset of D_N is recursive.

Proof. We must show that, for each $k \in D_N$, the function $S_k : D_N \to D_N$ given by

$$S_k(n) = \begin{cases} 0 & \text{if } n = k \\ 1 & \text{otherwise} \end{cases}$$

is recursive. We do this by induction on k.

Base step:

$$S_0(n) = \begin{cases} 0 & \text{if } n = 0 \\ 1 & \text{if } n \neq 0. \end{cases}$$

S_0 is just the function sg, so it is recursive.

Induction step: Suppose that $k > 0$ and that S_{k-1} is recursive.

$$S_k(n) = \begin{cases} 0 & \text{if } n = k \\ 1 & \text{if } n \neq k, \end{cases}$$

so we have

$$S_k(0) = 1,$$

and

$$S_k(n+1) = S_{k-1}(n) \text{ for every } n \in D_N.$$

Thus S_k is obtained by recursion from S_{k-1} and is therefore recursive.

Hence, by induction, S_k is recursive for each $k \in D_N$, so the result is proved.

Example 6.25

(a) Every finite set is recursive. This is a consequence of Proposition 6.24 and Corollary 6.23, since a finite set can be written as a finite union of singleton sets.

(b) Define $p : D_N \to D_N$ by

$$p(n) = n\text{th odd prime number, if } n > 0,$$

and

$$p(0) = 2.$$

Then p is a recursive function. This may seem surprising, for it is known that p has no simple algebraic expression. Nevertheless it can be shown from the definition (with a number of intermediate stages) that p falls into the class of recursive functions.

(c) According to elementary number theory, every natural number can be expressed uniquely as a product of powers of prime numbers. Define, for each $i \in D_N$, a function e_i by:

$$e_i(n) = \begin{cases} \text{exponent of the prime } p(i) \text{ in the expression of } n \text{ as} \\ \text{a product of powers of primes, if } p(i) \text{ occurs in that} \\ \text{expression, 0 otherwise.} \end{cases}$$

Then for each i, e_i is a recursive function.

(d) The function $d: D_N \to D_N$, where

$$d(m, n) = \text{the greatest common divisor of } m \text{ and } n,$$

is recursive.

▷ From simple basic functions and rules, complicated recursive functions can be built up, and clearly there is no limit to the complications in the application of rules I, II, and III. The proof of Proposition 6.12, which may be found in Mendelson, is quite lengthy and depends considerably on proofs that particular functions and relations are recursive and on results about how recursive functions may be combined to give other recursive functions and relations.

As a corollary to Propositions 6.10 and 6.12, we can obtain the rather unhelpful result that not all functions on D_N are recursive. Indeed this can be shown more directly by a countability argument without using these propositions.

The set of recursive functions is countable, and this fact enables us to construct a function which is not recursive. Consider an enumeration $f_1, f_2, \ldots$ of all one-place recursive functions. Define a function $g: D_N^2 \to D_N$ by

$$g(m, n) = f_m(n).$$

Then g is a non-recursive function. For suppose that g is recursive. Define h by

$$h(m) = g(m, m) + 1 \qquad (m \in D_N).$$

h is recursive since g is recursive, and so h is identical with f_k, for some

k. For this k, we have

$$h(k)=f_k(k)=g(k,k)+1=f_k(k)+1.$$

From this contradiction we deduce that g is not recursive.

Exercises

7　Show that the following functions are recursive.
(a)　　$e: D_N^2 \to D_N$, given by $e(m,n)=m^n$.
(b)　　min: $D_N \to D_N$, given by

$$\min(m,n)=\begin{cases} m & \text{if } m\leq n \\ n & \text{if } m>n. \end{cases}$$

(c)　　$q: D_N^2 \to D_N$, given by

$$q(m,n)=\begin{cases} \text{quotient on division of } n \text{ by } m \text{ if } m\neq 0 \\ 0 \quad \text{if } m=0. \end{cases}$$

8　Let R be a recursive $(k+1)$-place relation such that for every $n_1,\ldots,n_k \in D_N$ there is at least one $n_{k+1}\in D_N$ such that $R(n_1,\ldots,n_k,n_{k+1})$ holds. Prove that the function f defined by $f(n_1,\ldots,n_k)=\mu x[R(n_1,\ldots,n_k,x)]$ $(n_1,\ldots,n_k \in D_N)$ is recursive.

9　Let $e_2: D_N \to D_N$ be given by:

$e_2(n)=$ exponent of 2 in the expression of n as a product of powers of primes.

$(e_2(n)=0$ if 2 does not occur in that expression.)
Prove that e_2 is a recursive function.

10　Repeat Exercise 9, where e_2 is replaced by the function e_k whose value is the exponent of the prime p_k (k fixed).

11　Let f and g be recursive functions. Show that the function h given by

$$h(x)=f(x)^{g(x)} \qquad (x\in D_N)$$

is recursive.

12　Prove that, for any $n>1$, if $R_1,\ldots,R_n$ are k-place recursive relations, then $R_1 \wedge \ldots \wedge R_n$ and $R_1 \vee \ldots \vee R_n$ are recursive relations.

13　Prove that, for $n>1$, if $A_1,\ldots,A_n$ are recursive sets, then $A_1 \cap \ldots \cap A_n$ and $A_1 \cup \ldots \cup A_n$ are recursive sets.

14　Let R be a binary relation on D_N and let k be a fixed natural number. Define one–place relations (i.e. predicates) on D_N as follows:

$S(n)$ holds if and only if there is $m<k$ such that $R(m,n)$ holds.
$T(n)$ holds if and only if $R(m,n)$ holds for every $m<k$.

Prove that if R is a recursive relation, so are S and T. Can your proof be modified to apply to the relations with the condition '$<k$' removed?

6.4 Gödel numbers

One of the essential techniques which Gödel used in the proof of the incompleteness of $\mathcal{N}$ has become a standard procedure in logic and other fields. The idea is that of *code numbers*. Generally, information may be presented in the English language or any other language, or in an abstract symbolic language. In order to discuss, transmit or process this information it may be convenient (or even essential) to put it in numerical form. For example, when information is to be processed by a machine, very often it is converted first into some kind of numerical form. To exemplify in a crude way how this might be done, the words in a particular English dictionary could be numbered in order from (say) 20 upwards. The standard punctuation marks could be assigned numbers less than 20. A given English sentence could then be written as a sequence of numbers.

What Gödel did was to give a similar construction for the first order language $\mathcal{L}$ (still arbitrary and unspecified), so that each symbol, term, *wf.* and sequence of *wfs.* of $\mathcal{L}$ is assigned a code number in such a way that from any given code number, the corresponding expression of $\mathcal{L}$ is easily recoverable. There are different ways of doing this, and we describe one.

First, define a function g on the set of symbols of $\mathcal{L}$ thus:

$$g(() = 3,$$

$$g()) = 5,$$

$$g(,) = 7,$$

$$g(\sim) = 9,$$

$$g(\rightarrow) = 11,$$

$$g(\forall) = 13,$$

$$g(x_k) = 7 + 8k \quad \text{for } k = 1, 2, \ldots,$$

$$g(a_k) = 9 + 8k \quad \text{for } k = 1, 2, \ldots,$$

$$g(f_k^n) = 11 + 8 \times (2^n \times 3^k) \quad \text{for } n = 1, 2, \ldots; k = 1, 2, \ldots,$$

$$g(A_k^n) = 13 + 8 \times (2^n \times 3^k) \quad \text{for } n = 1, 2, \ldots; k = 1, 2, \ldots$$

Notice that each symbol is assigned a different odd positive integer. Notice also that any given odd positive integer (if it corresponds to any symbol at all) can be easily broken down in order to find out which symbol it comes from.

Example 6.26

(*a*) Find which symbol (if any) corresponds to the number 587.

If 587 corresponds to any symbol, it must be one of the x_k, a_k, f_k^n or A_k^n, so we must first find the remainder on division by 8.

$$587 = 8 \times 73 + 3 = 8 \times 72 \times 11,$$

and $72 = 2^3 \times 3^2$, so 587 corresponds to the function letter f_2^3.

(*b*) Show that 333 does not correspond to any symbol of $\mathscr{L}$.

$$333 = 8 \times 41 + 5 = 8 \times 40 + 13,$$

but $40 = 2^3 \times 5$, which is not of the form $2^n \times 3^k$, so 333 does not correspond to any symbol of $\mathscr{L}$.

$\triangleright$ Of course, in a particular language $\mathscr{L}$, not all the symbols will occur, so not every code number will be used.

Now a term or *wf.* in $\mathscr{L}$ is a string of symbols of $\mathscr{L}$, and we can assign numbers to such strings in the following way. If $u_1, \ldots, u_k$ are symbols of $\mathscr{L}$ let us denote by $u_1 u_2 \ldots u_k$ the string of symbols (which may or may not be a term or *wf.* of $\mathscr{L}$) and define

$$g(u_1 u_2 \ldots u_k) = 2^{g(u_1)} \times 3^{g(u_2)} \times \ldots \times p_k^{g(u_k)},$$

where, for each $i > 0$, p_i denotes the ith odd prime number, and $p_0 = 2$. Since each number can be expressed uniquely as a product of powers of primes, there is an obvious procedure for finding the string of symbols (if there is one) corresponding to a given number. Also different strings of symbols must necessarily have different code numbers.

Example 6.27

(*a*) $\quad g(f_1^1(x_1)) = 2^{g(f_1^1)} \times 3^{g(()} \times 5^{g(x_1)} \times 7^{g())}$

$\qquad\qquad = 2^{59} \times 3^3 \times 5^{15} \times 7^5.$

(*b*) $\quad g((A_1^2(x_1, x_2) \to A_1^1(x_1)))$

$\qquad = 2^{g(()} \times 3^{g(A_1^2)} \times 5^{g(()} \times 7^{g(x_1)} \times 11^{g(,)} \times 13^{g(x_2)} \times 17^{g())}$

$\qquad\quad \times 19^{g(\to)} \times 23^{g(A_1^1)} \times 29^{g(()} \times 31^{g(x_1)} \times 37^{g())} \times 41^{g())}$

$\qquad = 2^3 \times 3^{109} \times 5^3 \times 7^{15} \times 11^7 \times 13^{23} \times 17^5 \times 19^{11} \times 23^{61} \times 29^3$

$\qquad\quad \times 31^{15} \times 37^5 \times 41^5.$

(*c*) Any number in which a prime occurs to an even power or in which there are gaps in the sequence of primes occurring, cannot correspond to any string of symbols.

Remark 6.28

Code numbers of symbols are odd numbers. Code numbers of strings of symbols are even (since the prime number 2 always occurs with a non-zero exponent in the code number of a string). The two kinds of code numbers can thus be easily distinguished.

▷ Code numbers can be assigned to finite sequences of strings of symbols by a further extension of this process. Let $s_1, s_2, \ldots, s_r$ be strings of symbols of $\mathscr{L}$, and define

$$g(s_1, s_2, \ldots, s_r) = 2^{g(s_1)} \times 3^{g(s_2)} \times \ldots \times p_r^{g(s_r)}.$$

Notice that a given number cannot be the code number of a sequence in this way and at the same time be the code number of any individual string of symbols, because the exponent of 2 is even in the code number of a sequence and is odd in the code number of a string (by Remark 6.28).

We have now defined a function g from the set of all symbols, strings of symbols and finite sequences of strings of symbols of $\mathscr{L}$, with values in D_N. This function is one–one, but, as we have seen, it is not onto. It has been defined in such a way that there is an effective procedure (namely using the expression as a product of powers of primes) for calculating g^{-1} for any number in the range of g. The values of g we shall call Gödel numbers. Each term or *wf.* of $\mathscr{L}$ is a string of symbols, and so has a Gödel number. A proof or deduction in $K_{\mathscr{L}}$ is a finite sequence of strings of symbols, and so has a Gödel number.

Gödel's purpose in devising this coding system was in order to transform assertions about a formal system ($\mathscr{N}$ for example) into assertions about numbers, and then to express these assertions within the formal system. The assertions which can be made about a formal system concern *wfs.*, theorems and proofs. For example: 'The sequence $\mathscr{A}_1, \ldots, \mathscr{A}_k, \mathscr{A}$ is a proof in $\mathscr{N}$ of the *wf.* $\mathscr{A}$'. This assertion says that a certain relation holds between a finite sequence of *wfs.* and a particular *wf.* By means of Gödel numbers this gives rise to a relation on D_N, say *Pf*, defined by: $Pf(m, n)$ holds if and only if m is the Gödel number of a sequence of *wfs.* on $\mathscr{N}$ which constitutes a proof in $\mathscr{N}$ of the *wf.* whose Gödel number is n.

Other properties of, and assertions about, $\mathscr{N}$ give rise in a similar way to relations on D_N. It is now that the question of expressibility becomes important, for, continuing the above example, if the relation *Pf* were expressible in $\mathscr{N}$ there would be a *wf.* $\mathscr{P}(x_1, x_2)$ of $\mathscr{L}_N$ such that for every $m, n \in D_N$

$$\text{if } Pf(m, n) \text{ holds then } \underset{\mathscr{N}}{\vdash} \mathscr{P}(0^{(m)}, 0^{(n)}),$$

and

if $Pf(m, n)$ does not hold then $\underset{\scriptscriptstyle N}{\vdash} \sim\mathcal{P}(0^{(m)}, 0^{(n)})$.

To put it roughly, there would be a *wf.* $\mathcal{P}(x_1, x_2)$ which would decide *within the system* the 'metaquestion', for any sequence of *wfs.* $\mathcal{A}_1, \ldots, \mathcal{A}_k, \mathcal{A}$, whether it constitutes a proof in $\mathcal{N}$. What we are doing amounts to an attempt to use the system $\mathcal{N}$ (at least partially) as a metasystem, in the sense discussed in previous chapters, for itself. On the face of it, such a procedure seems dangerous and liable to lead to a contradiction, but we know that only recursive relations on D_N are expressible in $\mathcal{N}$, so we certainly cannot follow this procedure through for all relations on D_N, and the use of $\mathcal{N}$ as its own metasystem will necessarily be only partial. Because of this a contradiction can be avoided.

The next step in the proof of Gödel's theorem is to show that certain relations on D_N which arise in this way from considerations about *wfs.*, theorems and proofs are recursive and therefore expressible in $\mathcal{N}$. We shall omit the details, and merely list some such relations.

Proposition 6.29

The following relations on D_N are recursive, and therefore expressible in $\mathcal{N}$.

(i) *Wf*   *Wf*(n) holds if and only if n is the Gödel number of a *wf.* of $\mathcal{N}$.

(ii) *Lax*   *Lax*(n) holds if and only if n is the Gödel number of a logical axiom of $\mathcal{N}$.

(iii) *Prax*   *Prax*(n) holds if and only if n is the Gödel number of a proper axiom of $\mathcal{N}$.

(iv) *Prf*   *Prf*(n) holds if and only if n is the Gödel number of a proof in $\mathcal{N}$.

(v) *Pf*   *Pf*(m, n) holds if and only if m is the Gödel number of a proof of the *wf.* with Gödel number n.

(vi) *Subst*   *Subst*(m, n, p, q) holds if and only if m is the Gödel number of the result of substituting the term with Gödel number p for all free occurrences of the variable with Gödel number q in the expression with Gödel number n.

(vii) *W*   *W*(m, n) holds if and only if m is the Gödel number of a *wf.* $\mathcal{A}(x_1)$ in which x_1 occurs free, and n is the Gödel number of a proof in $\mathcal{N}$ of $\mathcal{A}(0^{(m)})$.

(viii) *D*   *D*(m, n) holds if and only if m is the Gödel number of a *wf.* $\mathcal{A}(x_1)$ in which x_1 occurs free, and n is the Gödel number of the *wf.* $\mathcal{A}(0^{(m)})$.

Exercises

15 Find the symbols of $\mathscr{L}$ (if any) corresponding to the following code numbers.
   (a)   65   (b) 299
   (c)   109   (d) 421.

16 Find the *wfs.* of $\mathscr{L}$ corresponding to the following code numbers.
   (a) $2^{61} \times 3^3 \times 5^{15} \times 7^5$;
   (b) $2^9 \times 3^{61} \times 5^3 \times 7^{15} \times 11^5$;
   (c) $2^3 \times 3^{13} \times 5^{15} \times 7^5 \times 11^{61} \times 13^3 \times 17^{15} \times 19^5$.

17 Every natural number corresponds to a unique sequence of natural numbers, through its expression as a product of powers of primes, e.g. the number $2^4 \times 3 \times 5^7 \times 11^2$ corresponds to the sequence 4, 1, 7, 0, 2. Two sequences can be combined by concatenation, i.e. extending one sequence by appending the elements of the other. For example, if s is the sequence 2, 3, 5 and t is the sequence 4, 7, 9, 10, then denote by $s * t$ the sequence 2, 3, 5, 4, 7, 9, 10. Define a function $f: D_N^2 \to D_N$ by:

$f(m, n) =$ the code number of $s * t$, where s and t are the sequences
whose code numbers are m and n.

Prove that f is recursive.

6.5 The incompleteness proof

The relation W defined in Proposition 6.29 is the key to the incompleteness proof, so an effort should be made to grasp its meaning. Note that it involves substitution of the term $0^{(m)}$ (which corresponds to the number m) into the *wf.* $\mathscr{A}(x_1)$, whose Gödel number is m.

W is expressible in $\mathscr{N}$, so there is a *wf.* $\mathscr{W}(x_1, x_2)$ in which only x_1 and x_2 occur free, such that

$$\text{if } W(m, n) \text{ holds then } \underset{\mathscr{N}}{\vdash} \mathscr{W}(0^{(m)}, 0^{(n)}),$$

and

$$\text{if } W(m, n) \text{ does not hold then } \underset{\mathscr{N}}{\vdash} \sim \mathscr{W}(0^{(m)}, 0^{(n)}).$$

Consider the *wf.*

$$(\forall x_2) \sim \mathscr{W}(x_1, x_2).$$

Let p be the Gödel number of this *wf.* and consider finally the *wf.* obtained by substituting $0^{(p)}$ for x_1, namely

$$(\forall x_2) \sim \mathscr{W}(0^{(p)}, x_2).$$

Denote this last *wf.* by $\mathcal{U}$. It will help at this stage to give a rough interpretation of $\mathcal{U}$, in order to see its significance. Firstly, W is the interpretation of $\mathcal{W}$. Thus $\mathcal{U}$ can be interpreted as:

'For every $n \in D_N$, $W(p, n)$ does not hold.'

Expanding this, it becomes:

'For every $n \in D_N$, it is not the case that p is the Gödel number of a *wf.* $\mathcal{A}(x_1)$ in which x_1 occurs free and n is the Gödel number of a proof in $\mathcal{N}$ of $\mathcal{A}(0^{(p)})$.'

Now p *is* the Gödel number of a *wf.* in which x_1 occurs free, namely $(\forall x_2) \sim \mathcal{W}(x_1, x_2)$, and if this *wf.* is denoted by $\mathcal{A}(x_1)$, then $\mathcal{A}(0^{(p)})$ is the *wf.* $\mathcal{U}$. Hence the interpretation of $\mathcal{U}$ is equivalent to:

'For every $n \in D_N$, n is not the Gödel number of a proof in $\mathcal{N}$ of the *wf.* $\mathcal{U}$.'

In a sense, therefore, the *wf.* $\mathcal{U}$ can be thought of as asserting its own unprovability.

If $\mathcal{N}$ were not consistent then it would trivially be complete, since every *wf.* would be a theorem. The incompleteness theorem will therefore require the hypothesis that $\mathcal{N}$ is consistent. In fact, Gödel's proof required a slightly stronger hypothesis, which we must now investigate.

Definition 6.30

A first order system S with the same language as $\mathcal{N}$ is ω-*consistent* if for no *wf.* $\mathcal{A}(x_1)$, in which x_1 occurs free, is $\sim(\forall x_1)\mathcal{A}(x_1)$ a theorem of S if $\mathcal{A}(0^{(n)})$ is a theorem of S for every $n \in D_N$.

As observed previously (see Remark 6.6), if $\mathcal{A}(0^{(n)})$ is a theorem for every n, $(\forall x_1)\mathcal{A}(x_1)$ need not necessarily be a theorem. ω-consistency asserts that if each $\mathcal{A}(0^{(n)})$ is a theorem then $\sim(\forall x_1)\mathcal{A}(x_1)$ is *not* a theorem, irrespective of whether $(\forall x_1)\mathcal{A}(x_1)$ is a theorem.

Proposition 6.31

Let S be a first order system with the same language as $\mathcal{N}$. If S is ω-consistent then S is consistent.

Proof. Let $\mathcal{A}(x_1)$ be any *wf.* for which $\mathcal{A}(0^{(n)})$ is a theorem of S for every n. For example, $\mathcal{A}(x_1)$ could be $x_1 = x_1$. Then, by ω-consistency, $\sim(\forall x_1)\mathcal{A}(x_1)$ is not a theorem of S. Thus S is consistent (since there is a *wf.* which is not a theorem).

Proposition 6.32 (Gödel's Incompleteness Theorem)

Under the assumption that $\mathcal{N}$ is ω-consistent, the wf. $\mathcal{U}$ is not a theorem of $\mathcal{N}$, nor is its negation. Hence, if $\mathcal{N}$ is ω-consistent then $\mathcal{N}$ is not complete.

Proof. First suppose that $\mathcal{U}$ is a theorem of $\mathcal{N}$ and let q be the Gödel number of a proof of $\mathcal{U}$ in $\mathcal{N}$. As before, let p be the Gödel number of $(\forall x_2) \sim \mathcal{W}(x_1, x_2)$. Thus $W(p, q)$ holds. W is expressible in $\mathcal{N}$ by $\mathcal{W}$, so we have

$$\vdash_{\mathcal{N}} \mathcal{W}(0^{(p)}, 0^{(q)}).$$

But $\vdash_{\mathcal{N}} \mathcal{U}$, i.e. $\vdash_{\mathcal{N}} (\forall x_2) \sim \mathcal{W}(0^{(p)}, x_2)$, and so $\vdash_{\mathcal{N}} \sim \mathcal{W}(0^{(p)}, 0^{(q)})$, using $(K5)$ and MP. This contradicts the consistency of $\mathcal{N}$, so $\mathcal{U}$ cannot be a theorem of $\mathcal{N}$.

$\mathcal{U}$ is not a theorem of $\mathcal{N}$, i.e. there is no proof in $\mathcal{N}$ of $\mathcal{U}$, so for no number q is q the Gödel number of a proof in $\mathcal{N}$ of $\mathcal{U}$, i.e. of $(\forall x_2) \sim \mathcal{W}(0^{(p)}, x_2)$. Hence $W(p, q)$ does not hold for any number q. Therefore

$$\vdash_{\mathcal{N}} \sim \mathcal{W}(0^{(p)}, 0^{(q)}) \quad \text{for every } q.$$

By ω-consistency, then,

$$\sim (\forall x_2) \sim \mathcal{W}(0^{(p)}, x_2)$$

is not a theorem of $\mathcal{N}$, i.e. $(\sim \mathcal{U})$ is not a theorem of $\mathcal{N}$.

Remark 6.33

We have explicitly stated the assumption of ω-consistency, although there is an obvious demonstration that $\mathcal{N}$ is ω-consistent, using the model N. As mentioned previously, however, arguments using models tend to involve assumptions, often about the consistency of other formal systems, and so tend to beg the question. Also, Proposition 6.32 can be generalised to cover other formal systems, extensions of $\mathcal{N}$, and certainly then the assumption of ω-consistency must be made, in the absence of any specific information.

▷ This chapter so far has been an outline version of the proof of Gödel's Incompleteness Theorem. This is included in an attempt to give some feel for the methods involved, and to facilitate an explanation of something of its significance. Let us now examine some consequences and generalisations, therefore.

Proposition 6.34 (Assuming that $\mathcal{N}$ is ω-consistent)
$\mathcal{N}$ contains a closed *wf.* which is true in the model N but which is not a theorem of $\mathcal{N}$.

Proof. The *wf.* $\mathcal{U}$ is a closed *wf.* Neither $\mathcal{U}$ nor $(\sim\mathcal{U})$ is a theorem of $\mathcal{N}$. However either $\mathcal{U}$ is true in N or $(\sim\mathcal{U})$ is true in N, since N is an interpretation.

▷ In fact, the assumption in this proposition can be weakened.

Proposition 6.35 (Assuming that $\mathcal{N}$ is consistent)
$\mathcal{N}$ contains a closed *wf.* which is true in the model N but is not a theorem of $\mathcal{N}$.

Proof. Modifications must be made to the proof of Proposition 6.32, which also holds with the weakened assumption, but the *wf.* $\mathcal{U}$ must be modified also. We omit details.

▷ $\mathcal{N}$ is not complete. Our first thought might now be: can we make $\mathcal{N}$ complete? Perhaps we chose an inadequate set of axioms of $\mathcal{N}$. Perhaps if we added the *wf.* $\mathcal{U}$ as a new axiom then the new system would be complete. Some thought about the procedures involved in this chapter should indicate that adding $\mathcal{U}$ as a new axiom will not help. Let $\mathcal{N}^+$ denote the system obtained from $\mathcal{N}$ by including $\mathcal{U}$ among the axioms. This change in the axioms does not affect the result that every recursive relation is expressible (though it might affect the converse). However the recursiveness of the relations *Prax* and *Pf* and others defined in terms of these may be affected. But the addition of a single axiom does not affect the recursiveness of the set of Gödel numbers of the axioms, for any singleton set is recursive and the union of two recursive sets is recursive. The relation *Prax* remains recursive, and in a similar way *Pf* and others, including *W*, can be seen to be recursive, though of course their definitions refer to $\mathcal{N}^+$ rather than $\mathcal{N}$. The same development as for $\mathcal{N}$ can then be worked through, leading to an incompleteness theorem involving a different *wf.* $\mathcal{U}'$.

By a more general argument along these lines, we obtain the following proposition.

Proposition 6.36
Let S be any extension of $\mathcal{N}$ such that the set of Gödel numbers of proper axioms of S is a recursive set. Then (provided that S is consistent) S is not complete.

Remark 6.37

The assumption about S is just the assumption that the relation $Prax_S$ defined by: $Prax_S(n)$ holds if and only if n is the Gödel number of a proper axiom of S, is recursive. It is this assumption which enables a proposition corresponding to Proposition 6.29 to be proved for S.

▷ It follows from this last proposition that $\mathcal{N}$ cannot be made complete by the inclusion of a set of additional axioms whose Gödel numbers constitute a finite or infinite recursive set.

As we have noted, the system $\mathcal{N}$ is deficient in one particular sense. Its axioms include only a weakened version of the Principle of Mathematical Induction. This is because of the use of a first order language. Could we avoid the difficulty of incompleteness by considering instead a second order system of arithmetic? In this book we have deliberately avoided the additional complication involved in second order systems, so any investigation of this question is beyond us. It is known, however, that second order arithmetic has the same property, i.e. a consistent second order system of arithmetic in which the set of (Gödel numbers of) proper axioms is recursive is not complete.

The consequences of Gödel's Theorem are still more far reaching than this. We have seen that the system of natural numbers can be defined within the formal system ZF. Any formal system of set theory, if it is adequate for set theory, will share this property, and indeed systems more restricted than ZF will share this property also. (An example of one which does *not* is our formal system of group theory, which is too restricted.)

Proposition 6.38

Any first order system which is sufficiently strong, whose set of (Gödel numbers of) proper axioms is recursive, and which is consistent, is not complete. (A system is sufficiently strong if the system of natural numbers can be defined in it as above and the axioms of arithmetic are then theorems.) In particular, if ZF is consistent, then it is not complete.

▷ Finally, let us consider what the observant reader may have already noticed, namely that there does exist an extension of $\mathcal{N}$ which is complete. The hypotheses of Gödel's Incompleteness Theorem include the requirement that the formal system concerned has a set of proper axioms whose Gödel numbers constitute a recursive set. This was a necessary supposition for the proof, since the proof involved demonstrations that certain relations were also recursive. That there is a first order system of arithmetic which is consistent and complete, and which therefore does not satisfy this requirement, can be seen using a procedure given in Chapter 4 as follows.

Consider the extension of $\mathcal{N}$ obtained by including as axioms all *wfs.* of $\mathcal{N}$ which are true in the model N. This extension is complete and is consistent, provided that $\mathcal{N}$ is consistent (see Corollary 4.49). By Proposition 6.36, then, the set of Gödel numbers of proper axioms of this extension cannot be recursive. It follows from this that the set of Gödel numbers of *wfs.* of $\mathcal{N}$ which are true in $\mathcal{N}$ is an example of a set which is not recursive.

Thus if we allow our sets of proper axioms to be non-recursive, we can have a first order system of arithmetic which is consistent and complete. The question which this brings up is: what is it about recursive sets which makes it worth while considering the result of Gödel's Incompleteness Theorem as significant? It is not true to say that no consistent formal system of arithmetic is complete. All we can say is that no such system whose set of (Gödel numbers of) axioms is recursive is complete. The answer lies in the ideas of computability, effectiveness and algorithm, and their relationship with the idea of recursiveness. We deal with these in the next chapter.

Exercise

18   Let the *wf.* $\mathcal{U}$ be as defined in the text. $\mathcal{U}$ is not a theorem of $\mathcal{N}$, so the extension of $\mathcal{N}$ obtained by including $(\sim\mathcal{U})$ as an additional axiom is consistent (presuming that $\mathcal{N}$ is consistent). Show that this extension is not ω-consistent.

7
Computability, unsolvability, undecidability

7.1 Algorithms and computability

At a world congress of mathematicians in 1900, Hilbert presented his celebrated list of outstanding unsolved problems of mathematics. One of these (now known as Hilbert's Tenth Problem) was the problem of finding a procedure for deciding whether any given polynomial Diophantine equation has a solution in integers. The solution to the problem (not found until very recently – see Davis [1]) was in terms which would have surprised Hilbert's 1900 audience (and possibly even Hilbert himself), because it does not consist of a set of instructions for the required procedure, but rather it is a proof that no such procedure can exist. Hilbert's Tenth Problem is an example of what is now popularly known as an 'unsolvable problem', or, to be more accurate, a recursively unsolvable problem. Investigations of this problem and others like it led mathematicians early this century to consider what was meant by the word 'procedure' as required by the problem. These considerations led to the ideas and applications described in this chapter.

In this context, another word with the same meaning as 'procedure' is the word 'algorithm' and we shall normally use the latter because the former has other meanings outside this context, which might lead to misunderstanding. The notion of algorithm is an intuitive one, not a mathematically precise one – we might define it in the following way.

Definition 7.1

An *algorithm* is an explicit effective set of instructions for a computing procedure (not necessarily numerical) which may be used to find the answers to any of a given class of questions.

When it is put in these terms questions of the existence of algorithms appropriate for different 'classes of questions' naturally arise. In Hilbert's Tenth Problem the class of questions is

{does there exist a solution in integers to the equation E?|
E is a polynomial Diophantine equation}.

Example 7.2

One may consider classes of questions of any sort, e.g.,

(a) {what is the value of $f(n)$? $|n \in D_N$}   (f a fixed function).
(b) {is n a member of the set A? $|n \in D_N$}   ($\mathscr{A}$ a fixed set).
(c) {is $\mathscr{A}$ a theorem of $\mathscr{N}$? $|\mathscr{A}$ a wf. of $\mathscr{N}$}.

Example 7.3

(a) {is 2 a factor of n? $|n \in D_N$}.

There is an algorithm which provides answers for such questions. Given any number n, find the remainder on division by 2 (using any of several well known elementary school procedures for division). If the remainder is 0, we give the answer Yes. If the remainder is 1, we give the answer No.

(b) {does n belong to the set of prime numbers? $|n \in D_N$}.

There is an algorithm which provides answers for such questions. Given any number n, there are standard methods for finding the remainders when n is divided by m, for each m $(1 < m < n)$. If none of these remainders is zero, then n is a prime number. If one or more of these remainders is zero, then n is not a prime number.

(c) {what is the value of $f(n)$? $|n \in D_N$}, where f is the function defined by $f(n) = 2n$ $(n \in D_N)$.

In this case elementary school arithmetic again provides an algorithm for computing the values of f.

(d) {what are the solutions amongst complex numbers to the equation E? $|E$ is a quadratic equation with integer coefficients}.

There is a well known algebraic formula which provides an algorithm in this case.

▷ It will be noted that a 'class of questions' is a very general notion in this context. Let us restrict ourselves for the moment to consideration of algorithms in relation to classes of questions of a particular nature, namely that exemplified by Example 7.3(c) above. In other words, let us consider the notion of 'function computable by algorithm'. Historically in the development of the subject it was this aspect which received earliest attention. Attempts were made independently by several researchers to make precise in a mathematical way the notion of algorithm and to characterise in a mathematical way the class of functions computable by algorithm. Of course whether any particular mathematical description of the notion corresponds exactly to the intuitive idea is not something which can be proved. As we shall see, however, there are good reasons for supposing that a certain mathematical description of algorithm is general enough to include all intuitive algorithms.

The descriptions given by the early researchers took different forms, which can be roughly categorised as follows.

(a)   abstract (precisely defined) computing machines,

(b)   formal constructions of computing procedures,

and

(c)   formal constructions yielding classes of functions.

The first two give characterisations of the notion of algorithm itself (in principle there is no difference between (a) and (b)). The last give descriptions of the class of functions computable by algorithm.

An example of (a) is *Turing machines*, which were devised by Turing in the 1930s. The idea is of a notional machine, which processes a notional tape on which is printed the input number (or numbers) in a coded form, by following certain predetermined rules of a simple and restricted nature, and produces on the tape at the end of the calculation the output number similarly in a coded form. Any algorithm for computing the values of a function is claimed to be translatable into instructions for such a machine.

An example of (b) is *Thue systems* which are purely formal systems in which sequences of symbols can be 'deduced' as consequences of other sequences of symbols by means of certain rules. Thus, given an input sequence, the rules allow this to be converted into an output sequence. (See Davis [2].)

An example of (c) is provided by the recursive functions. The basic functions and the rules are a formal construction for generating a class of functions.

All of these constructions have one thing in common, namely that they include partial functions. It is reasonable to say that a partial function is computable by algorithm if there is an algorithm which yields the value of the function whenever it is defined. In the Turing machine context this corresponds to allowing the rules for a machine to lead to a computation which never ends and so never gives an output. In the context of recursive functions, this corresponds to allowing the use of the least number operator in an unrestricted way, as indicated in Remark 6.13.

The crucial result is that all of the diverse characterisations of (partial) functions computable by algorithm led to the same class, namely the class of recursive partial functions.† This is something which is suscep-tible to proof, and has been proved. What is not susceptible to proof is whether this class of partial functions is exactly the class of functions

† The term 'partial recursive function' is standard in the literature, rather than 'recursive partial function', but we use the latter here for the sake of clarity.

computable by algorithm. However, in the light of the evidence and the lack of evidence to the contrary, we accept *Church's Thesis* which states:

> The class of partial functions computable by algorithm is identical with the class of recursive partial functions.

Remark 7.4

Acceptance of Church's Thesis amounts merely to a crystallisation of our intuitive idea of algorithm to correspond with the mathematical descriptions which have been given. There is no evidence that this is not a reasonable thing to do.

▷ Now, under the supposition of Church's Thesis, questions about the existence of algorithms become mathematically more manageable. For example, the question of whether there is an algorithm which gives the values of a particular function becomes the question of whether that function is recursive. In another context, the question of whether there is an algorithm which decides on membership of a given subset of D_N becomes the question of whether the characteristic function of that set is recursive, i.e. whether that set is recursive.

The usefulness of Church's Thesis lies in the fact that mathematical techniques can be used to show that a given function or set is (or is not) recursive, and thereby to demonstrate that an algorithm exists (or does not) for a particular class of questions. And conversely, it is often useful to deduce, having found a particular algorithm, that a corresponding set or function is recursive.

The reader may note that one half of Church's Thesis has more content than the other. Church's Thesis is equivalent to the conjunction of:

(i)   Every partial function computable by algorithm is a recursive partial function,

and

(ii)   Every recursive partial function is computable by algorithm.

The second of these assertions is amenable to demonstration, for an inductive 'proof' of it can be given, using a reasonable intuitive notion of algorithm. It proceeds by describing algorithms for computing each of the basic recursive functions (the base step of the induction), and then by showing how algorithms for computing particular functions can be used to construct algorithms for computing functions obtained from these by means of the rules I, II and III.

The first of the above assertions is the part of Church's Thesis which cannot be demonstrated. What has been proved is that for a number of different precise mathematical definitions of algorithm, all partial functions computable by that kind of algorithm are recursive.

It can be seen from all this that if techniques of recursiveness are to be used in discussion of existence of algorithms for particular classes of questions, Church's Thesis will be vital if the answer is that no algorithm exists. From the conclusion that a particular function or set is not recursive we can deduce that no algorithm exists only if we assume Church's Thesis.

Example 7.5

(a) Let f and g be one-place functions on D_N which are computable by algorithm. Then $f \circ g$ is computable by algorithm. For to compute $f \circ g\ (n)$, just compute $g(n)$ using the algorithm for g and then compute $f(g(n))$ using the algorithm for f. This can easily be generalised to cover application of rule II for the construction of recursive functions.

(b) Let $f: D_N \to D_N$ be defined by recursion from the function g thus:

$$f(0)\quad = k$$

$$f(n+1) = g(n, f(n)).$$

Suppose that g is computable by algorithm. We give an algorithm for computing $f(m)$, for any $m \in D_N$. If $m = 0$ then $f(m) = k$. If $m > 0$, compute $f(1) = g(0, f(0))$ using the algorithm for g, and then $f(2) = g(1, f(1))$, using the algorithm for g, and so on, until the value of $f(m)$ is obtained.

(c) A more specific case of (b) is the factorial function: $f(n) = n!$ $(n \in D_N)$. To calculate 10!, say, in practice what we do is to follow the procedure given in (b), computing successively $1!, 2!, 3!, \ldots, 9!, 10!$.

(d) Consider the function $h: D_N \to D_N$ given by $h(n) =$ the first non-zero digit in the decimal expansion of $1/n$. It is possible to prove that h is recursive directly from the definition, but it would be rather a complex procedure. Alternatively we can describe an algorithm which may be used to compute its values, and then, using Church's Thesis, deduce that it is recursive. The algorithm arises from the standard division procedure. First find the smallest value of k such that $n < 10^k$, and then find the quotient on division of 10^k by n.

(e) We know that the set of Gödel numbers of wfs. of $\mathcal{N}$ which are true in N is not recursive. It follows from Church's Thesis, then, that there is no algorithm for answering questions from the class

$$\{\text{is } \mathcal{A} \text{ true in } N? \,|\, \mathcal{A} \text{ a } wf. \text{ of } \mathcal{N}\}.$$

(f) Let A be the subset of D_N consisting of all numbers which are sums of two squares. A is recursive, but it would be rather difficult to prove directly that the function $f: D_N \to D_N$ given by

$$f(n) = \begin{cases} 0 & \text{if there exist } p, q \in D_N \text{ such that } n = p^2 + q^2 \\ 1 & \text{otherwise,} \end{cases}$$

is recursive. However, there is an algorithm which answers questions from the class $\{$is $n \in A$? $|n \in D_N\}$. We can describe it as follows. Given n, compute $p^2 + q^2$ for every pair of numbers p, q both less than n. If, for some pair p, q, $p^2 + q^2 = n$, then answer Yes. If for no pair p, q is $p^2 + q^2 = n$, then answer No. By Church's Thesis, then, A is a recursive set.

▷ What is the relevance of all this to the assumption in Proposition 6.36 that the set of Gödel numbers of proper axioms must be recursive? In the light of Church's Thesis, for any system S with this property there will be an algorithm which will answer questions from the set

$$\{\text{is } \mathscr{A} \text{ a proper axiom of } S? \ |\mathscr{A} \text{ a } wf. \text{ of } S\},$$

and, since the set of Gödel numbers of logical axioms of S (or any of our first order systems) is recursive, there will be an algorithm which will answer questions from the set

$$\{\text{is } \mathscr{A} \text{ an axiom of } S? \ |\mathscr{A} \text{ a } wf. \text{ of } S\}.$$

Moreover, for a system S which does not have this property there will be no such algorithm.

From the point of view which we took when the idea of formal system was introduced, i.e. attempting to use formal systems to reflect actual mathematical contexts and to make them precise, we can see that a system S for which there is no algorithm for deciding whether a given $wf.$ of S is an axiom of S is not satisfactory. It cannot help in deciding which statements of the mathematical context are true because there will be no effective procedure for deciding which statements correspond to $wfs.$ which are axioms, and there will be no effective procedure for deciding whether a given sequence of $wfs.$ is a proof. One purpose of the initial study of formal systems was to search for a formal procedure which would decide of any mathematical statement whether it was true by including amongst the provable $wfs.$ as many as possible of the true $wfs.$ A formal system in which the set of axioms is not recursive cannot be of use. A formal system whose set of axioms is recursive certainly meets the requirements for effective procedures for deciding what is an axiom and what is a proof. However, the incompleteness theorem says that even such a system (of arithmetic) cannot be of use either, because the set of theorems of the system does not include all the *wfs*. which are true (in the interpretation N).

Remark 7.6

The reader may care to consider the system $\mathcal{N}$ in which the set of (Gödel numbers of) proper axioms is recursive, and devise an effective procedure by which the set of theorems of $\mathcal{N}$ may be enumerated. (Hint: the relation *Pf* on pairs of Gödel numbers is recursive, and therefore there is an algorithm for deciding whether it holds of a given pair.) This shows that the set of (Gödel numbers of) theorems of $\mathcal{N}$ is 'effectively enumerable'. This brings us to a new notion.

Definition 7.7

A subset of D_N is *recursively enumerable* if it is the range of a recursive function, or if it is empty.

A set is recursively enumerable if there is a recursive function f such that $f(0), f(1), f(2), \ldots$ is a list of all members of the set (possibly with repetitions). Church's Thesis implies that 'recursively enumerable' is equivalent to 'effectively enumerable'.

▷ The immediate question we may now ask is: are 'recursive' and 'recursively enumerable' distinct notions? Is there a set which is recursively enumerable but not recursive, or vice versa? The following proof of a partial answer to this is a good illustration of the use which can be made of Church's Thesis.

Proposition 7.8

Every recursive set is recursively enumerable.

Proof. Let A be a recursive set, so that its characteristic function C_A is computable by algorithm. We describe an algorithm which enumerates the members of A. Calculate, in turn, the values of $C_A(0), C_A(1), \ldots$ and make a list of all numbers n such that $C_A(n) = 0$. By Church's Thesis, then, since A is effectively enumerable A is recursively enumerable.

▷ The converse of this proposition is false. There are several proofs of this, since all we need is a counterexample. We shall provide one by proving an important result about the system $\mathcal{N}$.

Definition 7.9

A formal system is *recursively undecidable* if the set of Gödel numbers of theorems of the system is not recursive.

Note that, using Church's Thesis, a formal system S is recursively undecidable if and only if there is no algorithm for answering questions from the set

$$\{\text{is } \mathscr{A} \text{ a theorem? } |\mathscr{A} \text{ a } wf. \text{ of } S\}.$$

▷ We shall prove later that $\mathscr{N}$ is recursively undecidable, after some preliminary work. Remark 7.6, along with Church's Thesis, indicates that the set of (Gödel numbers of) theorems of $\mathscr{N}$ is recursively enumerable. The recursive undecidability of $\mathscr{N}$ means that this set is not recursive.

Corollary 7.10

There exists a subset of D_N which is recursively enumerable but not recursive.

▷ The latter part of this chapter will concern recursive undecidability and the more general notion of recursive unsolvability. We already have an idea of what this latter notion is, but let us now make it precise, for future reference.

Definition 7.11

A class of questions is *recursively unsolvable* if there is no single algorithm which provides answers for all the questions in that class. (Note the implicit assumption of Church's Thesis in the use of the word 'recursively' in this definition.)

Thus a formal system S is recursively undecidable if and only if the class of questions

$$\{\text{is } \mathscr{A} \text{ a theorem of } S? \ |\mathscr{A} \text{ a } wf. \text{ of } S\}$$

is recursively unsolvable.

Exercises

1 Describe algorithms which provide answers to the following classes of questions.
   (a) $\{\text{is } n \text{ the Gödel number of a term in } \mathscr{N}? \ |n \in D_N\}$.
   (b) $\{\text{what is the greatest common divisor of } m \text{ and } n? \ |m, n \in D_N\}$.
   (c) $\{\text{is } n \text{ a perfect square? } |n \in D_N\}$.
   (d) $\{\text{is } \mathscr{A} \text{ a consequence in } L \text{ of the set } \Gamma? \ |\mathscr{A} \text{ a } wf. \text{ of } L\}$ (Γ a fixed finite set of wfs. of L).

(e) {what is the derivative of the function f? |f is a polynomial function of one real variable}.

(f) {what is the nth odd prime? |$n \in D_N$}.

2 Prove that for any subset A of D_N, if both A and its complement are recursively enumerable, then A is recursive. (Hint: use Church's Thesis.) Deduce that there is a subset of D_N which is neither recursive nor recursively enumerable.

3 Prove that if an infinite set A is recursively enumerable in increasing order (i.e. if there is a recursive function f such that $f(0), f(1), f(2), \ldots$ is a list of the members of A and $f(n) < f(n+1)$ for each $n \geq 0$), then A is recursive.

4 Show that the following functions are recursive.

(a) ϕ, where $\phi(n) =$ the number of positive integers p less than n such that p and n have no common factors ($n \in D_N$).

(b) f, where $f(m, n) =$ smallest integer greater than m/n, ($m, n \in D_N$).

(c) g, where

$$g(n) = \begin{cases} 0 \text{ if there is a sequence of at least } n \text{ 7s} \\ \quad \text{in the decimal expansion of } \pi, \\ 1 \text{ otherwise.} \quad (n \in D_N) \end{cases}$$

(d) q, where

$$q(m, n) = \begin{cases} \text{the Gödel number of } (\mathscr{A} \to \mathscr{B}) \text{ if } m \text{ and } n \text{ are} \\ \quad \text{the Gödel numbers of } wfs. \ \mathscr{A} \text{ and } \mathscr{B} \text{ of } \mathscr{N}, \\ 0 \text{ otherwise.} \quad (m, n \in D_N) \end{cases}$$

5 Give examples of

(a) A recursive set with a non-recursive subset.

(b) A non-recursive set with an infinite recursive subset.

6 Show that every infinite recursively enumerable set has an infinite recursive subset.

7.2 Turing machines

It is a valuable exercise actually to go into the details of one of the computational characterisations of the notion of algorithm. The Turing characterisation is the most useful and the most easily understood, and it can be applied directly to problems of decidability and solvability, as we shall see.

The reader should not be misled by the use of the word 'machine'. Turing machines are not actual working computing machines. They are abstract systems, precisely defined in a mathematical way so as to mirror computational procedures. The terminology that we use clearly indicates the 'machine' functions, and it is certainly true that actual machines can be built which follow the procedures of a 'notional' Turing machine.

Turing's purpose in describing his machines was to reduce computations to their barest essentials so as to describe in a simple way some basic procedures which are manifestly effective, and to which any effective procedure might be reduced. Let us now examine the technical details.

A Turing machine may be thought of as a black box through which passes a paper tape, which is divided into equal squares which may or may not have symbols printed on them. For a particular computation the machine will start with a finite amount of input information on the tape, i.e. with symbols printed on only finitely many tape squares. The machine processes the tape according to certain rules, and may or may not eventually come to a halt. If it does halt, then the output information is what remains on the tape. If it does not halt, then the computation is indeterminate and there is no output.

Before going further, there are two points raised above which require comment. The requirement that the input information be finite seems certainly to be a reasonable one. Indeed the reader may wonder why it is necessary to state it, for every existing paper tape is certainly finite in length and can consist of only finitely many equal squares. However it would be unreasonable to place a definite bound on the length of tape required for the input information. Also, as the machine is processing the tape, it may require more 'working space' than is provided on the original input tape, so we shall regard the tape as extendable indefinitely. Again, any computation will require only a finite amount of tape, but it would not be reasonable to put an absolute bound on the length of tape available, so we shall say that the tape is to be *potentially infinite*.

Besides placing no bounds on the tape, we shall place no bounds on the time that the machine requires for a particular computation. On a practical computer we would be forced to set a time limit, and, if no answer had been produced in that time, to give up and try to find a different program which would reduce the time required. However, for our abstract machines, it would be an artificial restriction to set an absolute limit on the number of steps or the time required to obtain an answer. All that we shall require is that if there is an answer, then the machine produces it in a finite time after a finite number of steps. Thus in the course of a computation we may not know (and in general we shall not know) whether that computation will terminate or not.

In order to study what such machines can do it is necessary to specify both the nature of the symbolic information which can appear on the tape and the ways in which the machine can process it.

A Turing machine has an *alphabet of tape symbols*, which may differ from machine to machine, but which is just a finite list of symbols. Each tape square may have printed on it at most one of these symbols at a time. Normally included in the list of symbols is the letter B, which will denote a blank square. The simplest Turing machine will have just two tape symbols, B and 1, say.

A Turing machine operates in the following way. At any given time the machine is 'reading' a single square of the tape. It may replace the symbol appearing in this square (if any) by another symbol, or print a symbol in it if it is blank, or leave the square unchanged, in which case it may shift its attention to the next square to the right or left on the tape. So we have:

Types of Operation:

   (a)   Print a symbol. (Printing a symbol includes first erasing the previous symbol.) Erasing a symbol, i.e. printing a B, is an operation of this type.
   (b)   Move one square left.
   (c)   Move one square right.

One *step* in the operation of the machine is a single operation of one of these types.

Next we must specify how the machine chooses, at each stage, which operation to perform. Its course of action is determined by the symbol which appears on the square being read and by the *internal state* of the machine. The machine is allowed to take on any of a finite number of internal states. In terms of actual computing machines the internal state may be thought of as the sum total of all the information stored in the machine at the given time. We are not concerned with the mechanics or electronics of storing information – we merely suppose that our black box has a finite number of different conditions which cause it to act in certain ways.

However, it is clear that provision must be made for the internal state of the Turing machine to change during a computation, so a single step in a computation must require specification of

   (1)   the present internal state of the machine,
   (2)   the contents of the square being read,
   (3)   the action taken by the machine, and
   (4)   the next internal state which the machine takes, in preparation for the next step of the computation.

Thus the internal state which the machine takes at a given time will be a consequence of the whole course of the previous computation, and in this sense it acts as a 'memory' for the machine.

Again, a comment is called for at this stage about the finiteness condition imposed on the number of internal states of a Turing machine. Actual digital computers do have only a finite number of different internal configurations, though admittedly this number is usually exceedingly large. However, for the same reasons as previously, it would be unreasonable to place any specific bound, even a very large one, on the number of states permitted in a Turing machine. We therefore require the number merely to be finite.

The most convenient way to specify the procedure which a Turing machine will follow is by means of a finite set of quadruples of the form

(state, tape symbol, action taken, new state taken).

In order to trace a computation through, at each stage it is necessary to note the internal state and the tape symbol being read, to search through the set of quadruples to find one which starts with this pair, and to follow the action given and take the new state given by this quadruple. This brings out a restriction which must be placed on our set of quadruples: it must be consistent, i.e. for each pair (state, symbol) there must be at most one quadruple which starts with this pair, so that the Turing machine will have a well-defined procedure to follow.

Since the number of states and the number of symbols are both finite, there is a limit on the number of quadruples for a particular machine. However, we shall not require that every (state, symbol) pair appears at the beginning of some quadruple. Some combinations may never occur in computations. More importantly, we shall say that a Turing machine computation *terminates* when and only when the current (state, symbol) pair does not occur in any of the quadruples, so that the machine has no instruction as to how to proceed.

Thus a Turing machine may transform an input tape into an output tape. The way in which information is symbolised on the tape must depend on the nature of the information, and we shall see by means of examples how it can be done. In all of our examples we shall use symbols q_0, q_1, q_2, ... to denote internal states, and we shall make the convention that the machines start in state q_0 reading the leftmost non-blank square on the tape. (Some convention of this kind is clearly necessary – there is no particular significance to this one.)

Example 7.12

$$\{(q_0 \ 1 \ B \ q_1), \ (q_1 \ B \ R \ q_0)\}$$

is the set of quadruples for a Turing machine with states q_0 and q_1 and alphabet of symbols B and 1. The quadruples are self explanatory except perhaps for the 'action taken' symbol. The third symbol in the quadruple can be L (for 'move left'), R (for 'move right') or just the symbol which is to replace the symbol being read. Observe the operation of this machine when the input tape contains a finite sequence of 1s, e.g.,

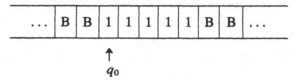

The machine starts in state q_0 reading the leftmost 1. It prints a B (erases the 1) and goes to state q_1.

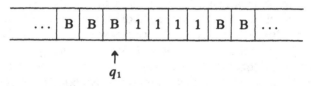

The machine in state q_1 reading a B moves one square to the right and goes to state q_0.

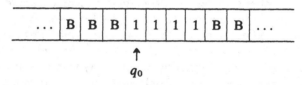

Now, just as above, the 1 on the square being read is replaced by a B and the machine goes to state q_1. Thus the machine will proceed, moving to the right replacing each 1 by a B, until it reaches the situation

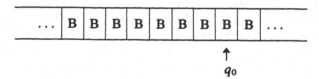

Now there is no quadruple to guide the machine in this situation, so the computation stops. This machine deletes a sequence of 1s and then stops.

Example 7.13

$$\{(q_0\ 1\ R\ q_1),\quad (q_1\ 1\ R\ q_0),\quad (q_1\ B\ R\ q_2),\quad (q_2\ B\ 1\ q_2)\}$$

is the set of quadruples for a Turing machine which decides whether a number is odd or even, in the following way. A number may be input in the form of a sequence of 1s on the tape. The machine starts in state q_0 reading the leftmost 1. E.g.,

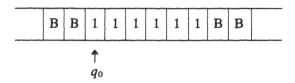

It proceeds as follows:

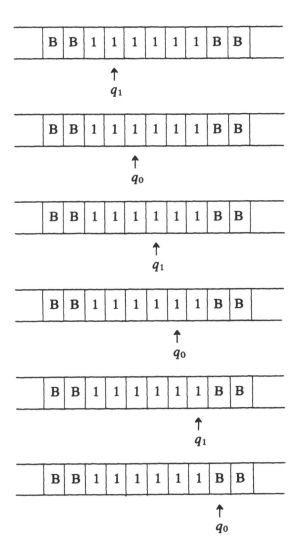

and halts there, since there is no quadruple starting with q_0B, and it will act similarly for any even input number. If the input number had been odd, we would have reached the situation (input 5)

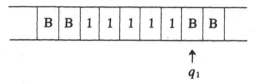

$\uparrow$
q_1

and the computation would proceed thus:

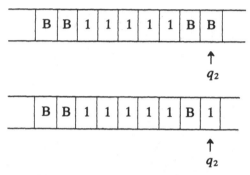

$\uparrow$
q_2

$\uparrow$
q_2

and halts here. This machine will therefore print a 1 after a space of one square if and only if the input number is odd.

Example 7.14

$$\{(q_0\ 1\ X\ q_0),\ \ (q_0\ X\ R\ q_0),\ \ (q_0\ 0\ Y\ q_0),\ \ (q_0\ Y\ R\ q_0)\}$$

is the set of quadruples for a Turing machine which, when a sequence of 0s and 1s is input, will translate it into a sequence of Xs and Ys, e.g. 1 0 1 0 0 1 is translated into X Y X Y Y X. Question: does this machine halt when this translation is completed?

Example 7.15

Addition is almost trivial for a Turing machine. If m and n are input on the tape as sequences of 1s separated by a letter A, the machine with quadruples

$$\{(q_0\ 1\ B\ q_0),\ \ (q_0\ B\ R\ q_1),\ \ (q_1\ 1\ R\ q_1),\ \ (q_1\ A\ 1\ q_2)\}$$

will delete the leftmost 1 and replace the A by a 1 and then stop. When it stops the number $m+n$ is left on the tape as a sequence of $(m+n)$1s.

Example 7.16

A Turing machine can reproduce the contents of the input tape. For example, if the input tape is blank except for a finite sequence of 0s and 1s, such a machine is specified by the set of quadruples:

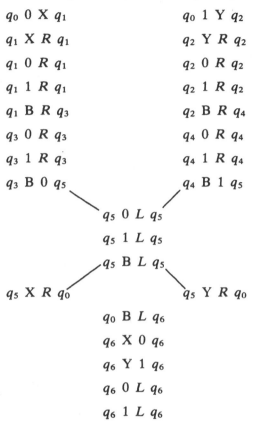

q_0 0 X q_1 q_0 1 Y q_2

q_1 X R q_1 q_2 Y R q_2

q_1 0 R q_1 q_2 0 R q_2

q_1 1 R q_1 q_2 1 R q_2

q_1 B R q_3 q_2 B R q_4

q_3 0 R q_3 q_4 0 R q_4

q_3 1 R q_3 q_4 1 R q_4

q_3 B 0 q_5 q_4 B 1 q_5

q_5 0 L q_5

q_5 1 L q_5

q_5 B L q_5

q_5 X R q_0 q_5 Y R q_0

q_0 B L q_6

q_6 X 0 q_6

q_6 Y 1 q_6

q_6 0 L q_6

q_6 1 L q_6

This example is clearly more complicated than the previous ones, and the quadruples have been laid out in this way in order to make clearer the machine procedure. Let us follow through a simple example.

Suppose that the input tape is blank except for ⎡ 0 1 ⎤ and that the machine starts in state q_0, reading the leftmost non-blank square. Then the machine will use first the left hand column of quadruples, and the tape will proceed thus:

| B | 0 | 1 | B | | B | X | 1 | B | | B | X | 1 | B |

  ↑ q_0 ↑ q_1 ↑ q_1

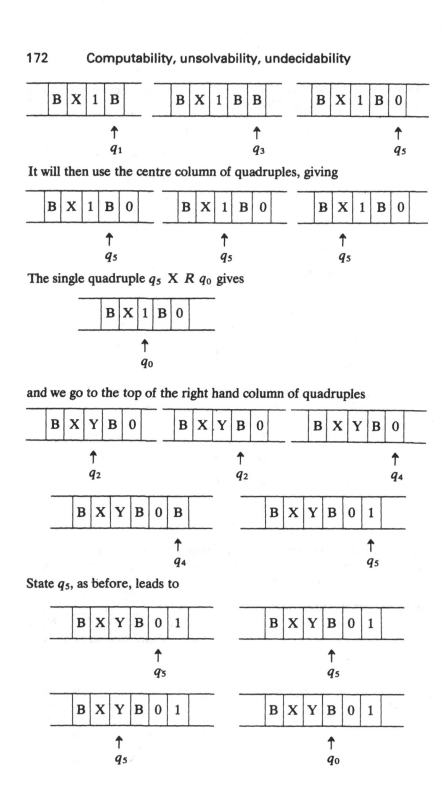

It will then use the centre column of quadruples, giving

The single quadruple q_5 X R q_0 gives

and we go to the top of the right hand column of quadruples

State q_5, as before, leads to

Now, state q_0 reading B uses the bottom centre column of quadruples, to give

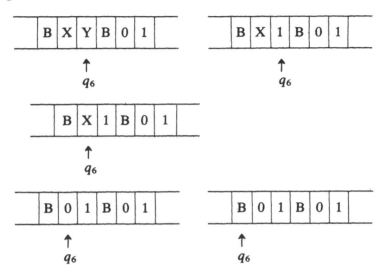

and here the machine halts, since there is no quadruple which starts with q_6 B.

▷ This last example illustrates some of the elementary processes of searching, copying, memorising and repeating which a Turing machine can use in carrying out more complex procedures, and it also illustrates how instructions (i.e. quadruples) can be found for a particular Turing machine by analysing the procedure which it is required to follow into a sequence of these elementary processes. Note that the size of the alphabet of tape symbols has a direct effect on the number of quadruples required. In particular, a machine which would duplicate a string of 1s requires the same sort of procedures as in the above example, but would need many fewer quadruples.

Example 7.17

In this example we shall not go into the same amount of detail as in the previous one, but the reader is recommended to follow through at least one actual machine computation in order to observe the combination of elementary processes involved and in particular the way in which the copying procedure of the previous example is used. The following quadruples specify a Turing machine which multiplies two numbers m and n when input in the form of sequences of m and n 1s, separated by a blank square.

q_0 1 X q_1	q_5 B L q_5
q_1 X R q_1	q_5 Y 1 q_6
q_1 1 R q_1	q_6 1 R q_2
q_1 B R q_2	q_2 B L q_7
q_2 1 Y q_3	q_7 1 L q_7
q_3 Y R q_3	q_7 B L q_7
q_3 1 R q_3	q_7 X B q_8
q_3 B R q_4	q_8 B R q_0
q_4 B 1 q_5	q_0 B R q_9
q_4 1 R q_4	q_9 1 B q_{10}
q_5 q L q_5	q_{10} B R q_9

This machine, with input tape, for example

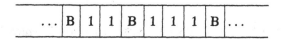

when started in state q_0 reading the leftmost 1, will eventually halt in state q_9 with the tape thus:

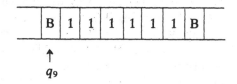

This machine, though designed for a more complicated procedure than the previous example, nonetheless has fewer quadruples. This is because the alphabet of tape symbols is here more restricted.

Remark 7.18

At any stage of a Turing machine computation, only a finite part of the tape is non-blank. We can therefore represent the condition of the machine by an 'instantaneous description', for example such as

$$1 \ 1 \ B \ 1 \ B \ q_2 \ B \ 1 \ B \ B \ B \ 1.$$

This includes the entire non-blank section of the tape and the square being read by the machine. The state symbol is included in the sequence, and is placed immediately to the left of the symbol being read. It is important to note that an instantaneous description need not contain *only* the non-blank part of the tape. For example,

$$q_5 \; B \; B \; B \; B \; 1 \; 1 \; B \; 1 \quad \text{and} \quad 1 \; B \; 1 \; 1 \; B \; B \; B \; q_3 \; B$$

are proper instantaneous descriptions where the machine is reading a square on a blank part of the tape. However, we never include any more blanks than are necessary to include the square being read.

Remark 7.19

A Turing machine is specified by its list of quadruples. By a coding system similar to the Gödel numbering of Chapter 6, it is possible to assign code numbers first to the quadruples themselves and second to finite lists of quadruples. This is because (we may suppose) all possible state symbols and tape symbols constitute a countable set, and each Turing machine uses only finitely many. To any Turing machine we can therefore assign a code number, and if the method of description (regarding initial state and how inputs and outputs are coded) is standardised, the code numbers will be such that all the relevant information about the corresponding Turing machines will be effectively recoverable. Just as previously, we choose the code numbers so that different code numbers correspond to different machines. Also it will be possible to decide effectively, given any natural number, whether it is the code number of a Turing machine. We shall not prove this, because it depends on the detail of the system of code numbers chosen, but it yields an important result.

Proposition 7.20

The collection of (code numbers of) Turing machines is effectively enumerable.

Proof. List all natural numbers, and as they appear delete all numbers which are not code numbers of Turing machines.

▷ This effective enumeration enables us to associate the list of Turing machines effectively with the set of all natural numbers. Each Turing machine will correspond to the number given by its position in the list. We can therefore take the set of code numbers of Turing machines to be the whole set of natural numbers. We have the following proposition.

Proposition 7.21 (The Enumeration Theorem)

The set of all Turing machines may be listed $T_0, T_1, T_2, \ldots$ in such a way that each suffix determines effectively and completely the instructions for the corresponding machine.

▷ Our examples of Turing machines have shown that they can perform various kinds of computation. We are interested in one particular kind of computation, namely computation of the values of functions. Now *any* Turing machine can be said to compute the values of a function provided that we specify the way in which the contents of the input and output tapes are interpreted as representing elements of the domain and range of the function. Our functions will be number theoretic functions, i.e. whose domains and ranges consist of natural numbers. An element n of D_N may be input to any Turing machine as a sequence of n 1s on an otherwise blank tape. When and if the machine halts subsequently, the output number may be taken to be the number of non-blank squares on the tape at that time. There is nothing special about these conventions. There is no reason why we should choose these rather than others, but it is important to be specific at this point, because we require to associate with each Turing machine a unique (partial) function. Note that these conventions yield functions of one variable. If we considered the computations of Turing machines when their input tapes have a pair of numbers represented on them (for example as described in Example 7.15), then the machines would be associated with functions of two variables. Clearly we can treat larger numbers of variables similarly.

Remark 7.22

(a) It is known that for any number theoretic function f which is computable by Turing machine, there is a Turing machine which computes the values of f and whose alphabet of tape symbols is {B, 1}. We shall not prove this, but it is a consequence of the fact that the alphabet of any Turing machine is finite, and its symbols can be encoded as numbers in binary notation, using symbol B in place of 0.

(b) It is known also that for any number theoretic function f which is computable by Turing machine there is a Turing machine which computes f and which has only two internal states. Again we shall not prove this but we should observe that this reduction in the number of states is possible only by considerably enlarging the alphabet of tape symbols. Either the number of states may be reduced to two, or the number of symbols may be, but not in general both. (See Minsky, Ch. 6.)

Definition 7.23

A number theoretic (partial) function is *Turing computable* if there is a Turing machine which computes its values, under the above specified conventions regarding input and output.

We have no information as yet regarding the possibility of a single function corresponding in this way to more than one Turing machine, but we can now be precise.

Proposition 7.24

For each Turing computable (partial) function f there are infinitely many Turing machines which compute the values of f.

Proof. Given f and a Turing machine T which computes its values, suppose that $q_0, \ldots, q_k$ are the internal states of T. Adjoin the quadruple $q_{k+1} \, 1 \, 1 \, q_{k+1}$ to obtain a new Turing machine T^1. T^1 computes the values of f, since the new quadruple has no effect on any of the computations (state q_{k+1} is never entered). Similarly we can generate a sequence $T^2, T^3, \ldots$ of Turing machines by adjoining the quadruples $q_{k+2} \, 1 \, 1 \, q_{k+2}, q_{k+3} \, 1 \, 1 \, q_{k+3}, \ldots$ successively. Each of these computes the values of f.

Remark. It is important to note the distinction which arises here between a function and a set of instructions for computing its values. The Turing machines referred to in the above proof are different, since they have different sets of quadruples, but the differences are inessential in that they do not affect the computations carried out.

▷ Turing postulated (in 1936) that he had succeeded in doing what he set out to do, namely to characterise in a mathematically precise way, by means of his machines, the class of functions computable by algorithm, and the following statement is now known as *Turing's Thesis*:

The class of Turing computable (partial) functions is the same as the class of (partial) functions computable by algorithm.

An equivalent way of stating this is to say that any algorithm (or set of instructions) for computing the values of a partial function f may be translated (effectively) into a set of quadruples for a Turing machine which computes the values of f.

As remarked previously, we cannot hope to prove Turing's Thesis, for it involves an intuitive notion (that of algorithm). However, a large amount of research work has led to many results which support Turing's Thesis. One of these is the following.

Proposition 7.25

A number theoretic (partial) function is Turing computable if and only if it is a recursive (partial) function.

Proof. The proof of this is highly technical, since it necessarily involves all the detail of the definitions of Turing machine and of recursive function. It is too lengthy to be included here, but the interested reader is referred to the book by Minsky.

▷ In the light of this, of course, Turing's Thesis is seen to be equivalent to Church's Thesis.

Proposition 7.26

There is an effective enumeration ϕ_0, ϕ_1, ϕ_2, ... of recursive partial functions of one variable, in which each recursive partial function occurs infinitely many times.

Proof. This is an immediate consequence of Propositions 7.21, 7.24 and 7.25.

▷ So much for the actual description of Turing machines and what they do. Let us now take things a stage further and discover where they lead us in regard to decision problems and solvability.

Consider the following algorithm: given any pair of numbers m, $n \in D_N$, enumerate the list T_0, T_1, T_2, ... of Turing machines until T_m is found, and follow through the computation of T_m applied to the input n. This is an algorithm for computing the values of a (partial) function of two variables. By Turing's Thesis, then, there is a Turing machine which computes the values of this function. This may be put in a proposition.

Proposition 7.27

There exists a *universal Turing machine*. That is, there is a Turing machine T, which when interpreted as computing the values of a function of two variables m and n, carries out the computation of the machine T_m with input n.

▷ The universal machine can thus carry out the procedures of any of the machines T_0, T_1, ... in computing functions of one variable. This is an indication of both the power and the limitations of Turing machines. The universal machine is apparently a complex and powerful object, for it embodies the capabilities of every one variable machine. On the other hand, the complexities of T_0, T_1, ... are limited by the complexity of the universal machine.

Consider the following algorithm: given any $n \in D_N$, find ϕ_n in the list of recursive partial functions, and follow the computation (using T_n) of $\phi_n(n)$. If and when a result is obtained, add 1 to that result. By Church's Thesis, this algorithm defines a recursive partial function, say ϕ_{k_0}. Now let us ask what is the result of the computation of the value of $\phi_{k_0}(k_0)$? Follow the algorithm, and we obtain the contradiction

$$\phi_{k_0}(k_0) = \phi_{k_0}(k_0) + 1.$$

We have missed a point, however, in obtaining this, and that is that the computation of $\phi_{k_0}(k_0)$ may not terminate. Indeed we can deduce from the above that it does not, for otherwise there is no way out of the contradiction.

The above illustrates how we can use the enumerations $T_0, T_1, \ldots$ and $\phi_0, \phi_1, \ldots$ in describing algorithms. This is a procedure which leads to interesting problems. It also can be used in the following way.

Proposition 7.28

There is no effective enumeration $f_0, f_1, \ldots$ of all recursive (total) functions of one variable.

Proof. Suppose the contrary, i.e. let $f_0, f_1, f_2, \ldots$ be an effective enumeration of all the recursive total functions of one variable (possibly with repetitions). Consider the following algorithm: given $n \in D_N$, enumerate $f_0, f_1, \ldots$ until f_n is obtained. Compute $f_n(n)$, and add 1. By Church's Thesis, this function h, with

$$h(n) = f_n(n) + 1 \quad \text{for each } n \in D_N,$$

is recursive, and it is total, since each f_n is total. So $h = f_k$ for some k. Thus

$$h(k) = f_k(k) = f_k(k) + 1.$$

This time there is no escape from the contradiction, and our result is proved.

▷ It would be convenient if for a given Turing machine computation we could tell in advance whether it will terminate. Thus the problem of deciding, given any pair $m, n \in D_N$, whether the Turing machine T_m will halt with input n, is a problem which has been examined. It is called the *Halting Problem* for Turing machines. Its importance is due in no small measure to the next result.

Proposition 7.29

The Halting Problem for Turing machines is unsolvable, i.e., there is no algorithm which provides answers to questions from the set

$$\{\text{does machine } T_m \text{ halt with input } n? \mid m, n \in D_N\}.$$

Proof. By Turing's Thesis it will be sufficient if we can show that the function $f: D_N \to D_N$ given by

$$f(n) = \begin{cases} 0 \text{ if } T_n \text{ halts on input } n \\ 1 \text{ if } T_n \text{ does not halt on input } n \end{cases}$$

is not Turing computable. For suppose that there is an algorithm for answering questions from the set given above. Then there is an algorithm for answering questions from the set

$$\{\text{does machine } T_n \text{ halt with input } n? \ | n \in D_N\},$$

and hence there is an algorithm for computing the values of the function f above. This cannot be the case if f is not Turing computable.

Let us concentrate on f, then, and suppose that it is Turing computable, and that the Turing machine T computes its values. Then, for input n, T will halt, with output 0 or 1 depending on whether T_n halts with input n or not. Let us now modify T to obtain a Turing machine T' such that for any n,

$$* \begin{cases} \text{if } T_n \text{ halts with input } n, T' \text{ does not halt with input } n, \text{ and} \\ \text{if } T_n \text{ does not halt with input } n, T' \text{ halts with input } n. \end{cases}$$

T' is obtained from T by including new states and quadruples which have the effect of appending, on to any computation by T which halts, a further search for a non-blank square on the output tape. If the search is successful, T' is to halt, otherwise T' is to continue searching indefinitely. The search can be successful (when T' has been started with input n) if and only if there is a 1 on the tape at this last stage, i.e. if and only if T_n does not halt with input n. (In detail, we could include two new states, q_α and q_β, say, and the following quadruples: $q_i \ S \ R \ q_\alpha$, for each pair $q_i \ S$ which do not occur as the first pair in any quadruple of T (S a tape symbol), and $q_\alpha \ B \ A \ q_\beta$, $q_\alpha \ A \ R \ q_\alpha$, $q_\beta \ A \ L \ q_\beta$, $q_\beta \ B \ A \ q_\alpha$.)

The machine T' must occur in the list $T_0, T_1, T_2, \ldots$. Say T' is T_{n_0}. Now let us ask the crucial question. Does T_{n_0} halt with input n_0? Return to * above. What it says is: T_{n_0} halts with input n_0 if and only if T_{n_0} does not halt with input n_0. This clear contradiction is sufficient to tell us that f cannot be Turing computable and that the result of the proposition is demonstrated.

▷ Consider now the following algorithm: given any $n \in D_N$, search in the list $T_0, T_1, T_2, \ldots$ until T_n is reached, and then follow the computation of the machine T_n with input n. If and when the computation halts, give output 1. By Turing's Thesis, the function whose values are computed by this algorithm is Turing computable. This function, ϕ, say, is given by

$$\phi(n) = \begin{cases} 1 \text{ if } T_n \text{ halts with input } n, \\ \text{undefined, otherwise.} \end{cases}$$

It is clear that ϕ is a partial function, for certainly there are Turing machines which halt for no input and others which halt for only some inputs. The domain of ϕ is an important set in this field of study, and is usually denoted by K.

$$K = \{n \in D_N : T_n \text{ halts with input } n\}.$$

Now it is a consequence of the proof of Proposition 7.29 that K is not a recursive set (under the assumption of Church's Thesis, of course), for if it were there would be an algorithm for answering questions of the form 'is $n \in K$?' for $n \in D_N$, i.e. questions from the set

$$\{\text{does machine } T_n \text{ halt with input } n? \mid n \in D_N\}.$$

Proposition 7.30

K is a recursively enumerable, non-recursive set.

Proof. K is not recursive, by the above. That K is effectively enumerable may be shown by giving an algorithm for enumerating it.

    Step 1. Follow one step in the computation of T_0 with input 0.

    Step 2. Follow one step in the computation of T_1 with input 1 and the second step in the computation of T_0 with input 0.

    Step 3. Follow one step in the computation of T_2 with input 2 and the second step in the computation of T_1 with input 1 and the third step in the computation of T_0 with input 0.

and so on. Whenever one of the machines T_i halts, put i in the enumeration of K and disregard the references to T_i in the subsequent steps of the algorithm. For each $n \in K$ there will come a step when T_n with input n is found to halt, and n is put into the enumeration of K. K is thus effectively enumerable, and, by Church's Thesis, recursively enumerable.

Proposition 7.31

For any Turing machine T, the domain of T, i.e. the set of all $n \in D_N$ for which T halts with input n, is a recursively enumerable set (which may also be recursive, of course).

Proof. The proof is very similar to the above proof, following the computations, this time of the same machine T, on different inputs, simultaneously making a list of all the input numbers for which T halts.

$\triangleright$ K is a concrete example of a non-recursive set which arises from consideration of Turing machines and the Enumeration Theorem.

There are many others. For example:

Example 7.32

(a) $\{n \in D_N: T_n$ halts for every input number$\}$ is not recursive, nor recursively enumerable.

(b) $\{n \in D_N: T_n$ halts for no input number$\}$ is not recursive, nor recursively enumerable.

(c)   For each fixed n_0, the set $\{n \in D_N: T_n$ halts with input $n_0\}$ is not recursive, but is recursively enumerable.

These examples also give us classes of questions which are recursively unsolvable, according to Definition 7.11.

(d) $\{$is $n \in K$? $|n \in D_N\}$

(e) $\{$does T_n halt for every input number? $|n \in D_N\}$

(f) $\{$does T_n halt for some input number? $|n \in D_N\}$

(g) $\{$does T_n halt for input n_0? $|n \in D_N\}$, where n_0 is any fixed number.

These classes of problems are all recursively unsolvable.

The method used to verify all these results is similar. It is to assume that the set is recursive (or that the class of problems is recursively solvable), and to deduce that the set K is recursive (or that some other known non-recursive set is recursive). This method may be viewed in two ways. First, it can be seen as a straight proof by contradiction. Second, it can be seen in a wider context as introducing a new idea, that of *reducibility*.

Definition 7.33

The set A is *reducible* to the set B if the existence of an algorithm for deciding membership of B would guarantee the existence of an algorithm for deciding membership of the set A.

Neither algorithm may exist, but it is of interest often to discover whether two sets are related in this way. An obvious example of such reducibility is the result that $D_N \backslash K$ is reducible to K, although neither set has an algorithm for deciding membership. These ideas will not concern us here, however, and the interested reader is referred to the book by Rogers for further information.

▷ This concludes our investigation of Turing machines. Now we shall proceed to use the ideas and techniques we have developed in a discussion of unsolvability and undecidability results.

Exercises

7   Modify the Turing machine of Example 7.12 so that it will delete every symbol occurring on the input tape. (As it stands the machine will delete 1s

until it comes to a B, when it will stop, so that if the input tape has a number of 1s scattered along it, not all will be deleted.)

8 Modify the Turing machine of Example 7.13 so that it halts with a completely blank tape if the input number is even and halts with a single 1 on the tape if the input number is odd.

9 Construct a Turing machine with alphabet of tape symbols {B, 1} which does not halt for any input tape.

10 Construct a Turing machine which when the input tape contains a single sequence of 1s will halt with the tape containing two sequences of 1s of that same length, separated by a symbol X.

11 Construct a Turing machine T which will compute the values of the partial function f, where

$$f(n) = \begin{cases} n & \text{if } n \text{ is odd} \\ \text{undefined} & \text{if } n \text{ is even.} \end{cases}$$

Modify T to obtain a Turing machine T' which behaves as T does if the input number is odd, but which halts with a blank tape if the input number is even. Give an example of a Turing computable partial function for which the above procedure is not possible.

12 Let T be a Turing machine whose set of quadruples contains no instruction to move left. Devise an effective procedure for deciding in advance for any input tape, whether T will eventually halt.

13 A Turing machine whose instructions permit it to move in only one direction along the tape is sometimes called a finite state machine. Show that the halting problem for finite state machines is solvable.

14 Prove that $\bar{K}$ is not recursively enumerable.

15 Let A be a recursively enumerable subset of $\bar{K}$ and suppose that for some Turing machine T_n, $A = \{x: T_n \text{ halts with input } x\}$. Show that $n \in \bar{K} \backslash A$.

16 Show that every recursively enumerable set is the domain of some Turing machine.

17 Show that each of the classes of problems in Example 7.32 is recursively unsolvable.

18 Let K_0 be the set $\{(m, n): \text{Turing machine } T_m \text{ halts with input } n\}$. Show that K is reducible to K_0, and that K_0 is reducible to K.

7.3 Word problems

One area where algebra and logic impinge on one another is in consideration of word problems for algebraic systems such as groups, semigroups and abelian groups. The case of semigroups is the simplest to describe, so we shall use it as an illustration.

Let $A = \{a_1, \ldots, a_k\}$ be a set of formal symbols. This set is to be regarded as an alphabet, and the *words* in this alphabet are just the finite strings of symbols from the alphabet (there are no restrictions on which strings are words). Denote by S_A the set of all words in the alphabet A.

Then S_A can be regarded as a semigroup under the operation of juxtaposition. For any two words in S_A a combined word can be formed just by appending the second after the first. This operation is automatically associative, so S_A is a semigroup. If we make the convention that the empty word (the word which consists of no symbols) is a member of S_A, then this word acts as an identity in the semigroup.

Now a word problem arises when a semigroup such as S_A above is modified by requiring it to satisfy one or more relations. We can illustrate this by an example.

Example 7.34

Consider S_A as above, and let us stipulate that the words a_1a_2 and a_2a_1 are to be identified with one another. From this we define an equivalence relation on S_A as follows. If P and Q are words, then $P a_1 a_2 Q \sim P a_2 a_1 Q$ and $P a_2 a_1 Q \sim P a_1 a_2 Q$, and for any words U and V, U is *equivalent* to V if there is a sequence of words $W_1, \ldots, W_n$ such that $U = W_1$, $W_1 \sim W_2, \ldots, W_{n-1} \sim W_n$ and $W_n = V$. It is easy to show that this is an equivalence relation and that the semigroup operation is well defined on equivalence classes. (The reader should note that $\sim$ is not an equivalence relation here – it is not transitive.) The set S_A^* of equivalence classes forms a semigroup. It is the semigroup with generators $a_1, \ldots, a_k$ and *relation* $a_1a_2 = a_2a_1$. The *word problem* for S_A^* is to decide, for any given words U and V, whether they are equivalent, i.e. whether they represent the same element of S_A^*.

▷ More generally, we can consider semigroups obtained from S_A by including a number of relations of the form $P_1 = Q_1$, $P_2 = Q_2, \ldots,$ $P_m = Q_m$, where $P_1, \ldots, P_m, Q_1, \ldots, Q_m$ are given words in the alphabet A. Here we define $\sim$ by

$$P P_i Q \sim P Q_i Q$$

for any words P, Q and $1 \le i \le m$, and we define the equivalence relation just as previously. The set of equivalence classes is called the semigroup with generators $a_1, \ldots, a_k$ and relations $P_i = Q_i$ $(1 \le i \le m)$. The word problem for this semigroup is the analogous problem to the previous one, but clearly will be more complicated the more relations there are.

Definition 7.35

A semigroup is *finitely presented* if it is obtained as above from a finite set of generators and a finite set of relations.

The word problem for a finitely presented semigroup is *recursively solvable* if there is an algorithm which will decide, given any pair of words, whether they are equivalent.

The *'Word Problem for Semigroups'* is: does there exist an algorithm which will decide, given any finitely presented semigroup and any pair of words in that semigroup, whether they are equivalent?

Example 7.36

Let $A = \{a_1, a_2\}$ and consider the semigroup S generated by A subject to the relation $a_1a_2 = a_2a_1$. We can see that this semigroup has recursively solvable word problem, by describing an algorithm as follows.

Every word involving just a_1 and a_2 can be rearranged in stages, using the relation $a_1a_2 = a_2a_1$ each time, until a word is obtained in which every occurrence of a_1 precedes every occurrence of a_2. This word is equivalent to the original word. (For example, $a_2a_1a_2a_1 \sim a_2a_1a_1a_2 \sim a_1a_2a_1a_2 \sim a_1a_1a_2a_2$.) For the word problem, given any two words P and Q, carry out this procedure on each. P and Q are equivalent if and only if the resulting words are identical.

$\triangleright$ As we observed previously, if an algorithm exists for a particular class of questions, finding it and describing it are usually possible without special methods or assumptions. On the other hand, in order to prove that no algorithm exists, virtually the only method open to us is through Church's Thesis. In fact there is a finitely presented semigroup whose word problem is not recursively solvable. Our proof of this relies entirely on the ideas and methods that we have developed in regard to Turing machines, so let us see first of all how we can relate the two notions, of Turing machine and of finitely presented semigroup.

We have seen how the condition of a Turing machine and its tape at a given moment can be specified by an 'instantaneous description', consisting of a string of symbols, e.g.

$$1 \; B \; 1 \; 1 \; B \; q_i \; 1 \; B \; 1, \quad 1 \; 1 \; 1 \; 1 \; B \; 1 \; B \; q_i \; B.$$

What we shall do is to consider these as words, and consider two words to be equivalent if one is transformed into the other by a sequence of Turing machine operations. There are certain complications, as we shall see, but from our Turing machine we take our alphabet A to include all the tape symbols and state symbols, and we take the semigroup relations as given by the Turing machine quadruples. For example, if $q_1 \; 1 \; B \; q_2$ is one of the quadruples, then we shall say that

$$P \; q_1 \; 1 \; Q \sim P \; q_2 \; B \; Q,$$

where P and Q are any strings of symbols from A. Of course, only some strings of symbols from A will actually have the form of an instantaneous description of the Turing machine – some strings will contain mor   han one state symbol, for example. We shall find that this does not matter. Let us now be specinc.

Let T be a Turing machine which halts if and only if the input α belongs to the set K (K is reeursively enumerable but not recursive). We shal suppose that T has the tape symbols 1 and B only, and that the number α is input as a sequence or α 1s. Let the internal states of T be denoted by $q_0, \ldots, q_n$, and let I be the set of quadruples of T. Take

$$A = \{q_0, q_1, \ldots, q_n, q, q', 1, B, h\},$$

and consider the set S_A of words in this alphabet. (The reason for the inclusion of the symbols q, q and h will be explained below.) Denote by $\mathscr{S}$ the fmitely presented semigroup with alphabet A and the following relations. (The letters X and Y are used to denote arbitrary tape symbols, 1 or B, throughout.)

1. q_i X $= q_j$ Y   if q_i X Y $q_j \in I.$

2. $\left. \begin{array}{l} q_i \text{ X Y} = \text{X } q_j \text{ Y} \\ q_i \text{ X } h = \text{X } q_j \text{ B } h \end{array} \right\}$ if q_i X R $q_j \in I.$

3. $\left. \begin{array}{l} \text{X } q_i \text{ Y} = q_j \text{ X Y} \\ h \, q_i \text{ Y} = h \, q_j \text{ B Y} \end{array} \right\}$ if q_i Y L $q_j \in I.$

4. q_i X $= q$ X   if there is no quadruple in I which starts with
          q_i X.

5. q X $= q.$

6.   X $q' = q'.$

7. q $h = q$ $h.$

We have included relations not only to correspond with the Turing machine instructions but also to govern the behaviour of q, q' and h. The machine's instantaneous description is to correspond with a word in S_A which is bounded by an h at each end. Thus the initial condition of the machine might be q_0 1 1 1 1. We shall consider this as being represented by the word h q_0 1 1 1 1 h. The hs delineate the significant part of the tape. The machine transforms its input tape in stages, and in corresponding stages our word is transformed, by applications of the relations, into equivalent words. The relations in 2 and 3 which involve h enable us to extend the word by the insertion of a B at one end when it is required (i.e. when the machine reads a square which is off the non-blank part of the tape). Note what happens when the machine halts. The machine does no more to its tape, but we can continue to transform our word using relations 4–7. Let us examine an example to see what happens. Suppose that the machine has halted and our word at that point was

$$h \ 1 \ B \ B \quad q_i \ 1 \ 1 \ h.$$

Then no quadruple in I starts with q_i 1, and relation 4 applies, and

$$h \ 1 \ B \ B \ q \ 1 \ 1 \ h$$

is an equivalent word. Relation 5 leads to equivalent words

$$h \ 1 \ B \ B \ q \ 1 \ h,$$

and

$$h \ 1 \ B \ B \ q \ h.$$

Relation 7 then yields

$$h \ 1 \ B \ B \ q' \ h,$$

and relation 6 gives equivalent words

$$h \ 1 \ B \ q' \ h,$$

$$h \ 1 \ q' \ h,$$

$$h \ q' \ h.$$

This final word is not dependent on the final contents of the tape. In fact, what we can say is that the machine T with input α halts if and only if the word $h \ q_0 \ \alpha \ h$ is equivalent to $h \ q' \ h$. It is this which enables us to show that $\mathcal{S}$ is a semigroup with recursively unsolvable word problem.

Proposition 7.37

There is a finitely presented semigroup whose word problem is recursively unsolvable

Proof. Let $\mathcal{S}$ be as described above. Suppose that there is an algorithm for deciding, for any pair of words of $\mathcal{S}$, whether they are equivalent. Then there is an algorithm for deciding membership of the set E, where

$$E = \{W \in S_A : W \text{ is equivalent to } h \ q' \ h\}.$$

Now T halts with input α if and only if $\alpha \in K$. Thus to decide membership of K we need only, given α, ask whether the word $h \ q_0 \ \alpha \ h$ is a member of E, using the result (not yet fully proved) that T halts with input α if and only if $h \ q_0 \ \alpha \ h$ is equivalent to $h \ q' \ h$. But there is no algorithm for deciding membership of K, since K is not recursive. Thus the assumption that $\mathcal{S}$ has recursively solvable word problem leads to a contradiction, and the proposition is proved subject to the following lemma.

Lemma. T halts with input α if and only if the words $h \; q_0 \; \alpha \; h$ and $h \; q' \; h$ are equivalent.

Proof. Suppose that $h \; q_0 \; \alpha \; h$ and $h \; q' \; h$ are equivalent. We show that T halts with input α (the implication the other way is demonstrated in the discussion preceding the proposition). A demonstration that the words are equivalent would consist of a transformation, using relations 1 to 7, starting with $h \; q_0 \; \alpha \; h$ and ending with $h \; q' \; h$. Now q' cannot be introduced other than by an application of 7. It follows that $h \; q_0 \; \alpha \; h$ is equivalent to a word in which the symbol q occurs. q can be introduced only by an application of 4. In a sequence of equivalent words leading from $h \; q_0 \; \alpha \; h$ to $h \; q' \; h$, then, consider the first occasion at which relation 4 is applied. Up till then each transformation must have corresponded to a Turing machine step (being an application of 1, 2, or 3). So we have

$$h \; q_0 \; \alpha \; h \text{ equivalent (say) to } h \; P \; q_i \; X \; Q \; h,$$

where $P \; q_i \; X \; Q$ is the instantaneous description of the Turing machine condition reached from the initial condition $q_0 \; \alpha$. Since 4 is applied at this point, $q_i \; X$ cannot occur in any quadruple of T, and so T halts when it reaches $P \; q_i \; X \; Q$. Thus T halts with input α.

Proposition 7.38

The word problem for semigroups is recursively unsolvable.

Proof. If there were an algorithm which would decide, for any finitely presented semigroup S and any words W_1 and W_2 of S whether they are equivalent, this algorithm could be used for the semigroup $\mathscr{S}$, contradicting the previous proposition.

$\triangleright$ We thus have our desired result for semigroups.

Definition 7.39

A *finitely presented group* is defined analogously with a finitely presented semigroup, except that formal inverses may occur in words (i.e. words are made up of the symbols $a_1, a_2, \ldots, a_k, a_1^{-1}, \ldots, a_k^{-1}$), and amongst the relations must always be $a_i a_i^{-1} = e$ and $a_i^{-1} a_i = e$ for $1 \le i \le k$, where e denotes the empty word.

Undecidability of formal systems   189

With this obviously more complicated situation, the word problem is more difficult, and the answer has been found only comparatively recently.

Proposition 7.40

(i) There is a finitely presented group whose word problem is recursively unsolvable.
(ii) The word problem for groups is recursively unsolvable.
(iii) The word problem for *abelian* groups is recursively *solvable*. (For a finitely presented abelian group, the set of relations must include $a_i a_j = a_j a_i$ for each pair of symbols a_i, a_j in the alphabet.)

Exercises

19   In each case below, describe an algorithm for solving the word problem of the finitely presented semigroup with the given generators and relations.
(a) $\{a_1, a_2\}$, $a_1 a_1 = a_1$.
(b) $\{a_1, a_2, a_3\}$, $a_2 a_2 = a_2$, $a_1 a_2 = a_3$.
(c) $\{a_1, a_2, a_3, a_4\}$, $a_1 a_1 = e$, $a_2 a_2 = e$, $a_3 a_3 = e$, $a_4 a_4 = e$.
20   Let G be the finitely presented group with generators $\{a_1, a_2, a_3\}$ and relations $a_1 a_2 = a_2 a_1$, $a_2 a_3 = a_3 a_2$, $a_3 a_1 = a_1 a_3$. Show that the word problem for G is recursively solvable.
21   Let S be a finitely presented semigroup with recursively unsolvable word problem. Let A be the set of generators of S. Show that S' has recursively unsolvable word problem, where S' is any finitely presented semigroup whose set of generators includes A and whose set of relations is the same as that of S.
22   Deduce from the result of Proposition 7.37 that there is a finitely presented semigroup with only two generators which has recursively unsolvable word problem.

7.4 Undecidability of formal systems

The reader will recall the result of Chapter 2 (Proposition 2.24) that the formal system L of propositional calculus is *decidable*. Of course at that time it meant less to us than it does now, for we can now prove:

Proposition 7.41

The set of Gödel numbers of theorems of L is a recursive set.

Proof. First note that for L we must define a new Gödel numbering, since L does not have the same symbols as our first order languages $\mathscr{L}$. This presents no difficulty, since L is simpler. For example, (and) can be assigned the numbers 3 and 5 as before, $\sim$ and $\rightarrow$ can be assigned 7 and

9 respectively, and each statement letter p_k can be assigned the number $9+2k$, for $k = 1, 2, \ldots$ *Wfs.* and sequences of *wfs.* are then assigned numbers just as before, using powers of primes.

For the proof of the proposition, we require an algorithm for answering questions from the set

$$\{\text{is } n \text{ the Gödel number of a theorem of } L? \mid n \in D_N\}.$$

Given $n \in D_N$, find the *wf.* of L corresponding to n (if there is one). If there is no such *wf.*, the answer required is No. If there is one, construct its truth table, and observe whether the *wf.* is a tautology or not.

▷ The question of recursive decidability or undecidability is indeed one which can be asked of any of our formal systems, because the idea of Gödel numbering applies to all of them. The question is then just one of the recursiveness or non-recursiveness of a particular subset of D_N.

Whether the system $K_{\mathscr{L}}$ of predicate calculus is recursively undecidable depends on the language $\mathscr{L}$. To take an extreme case, let $\mathscr{L}_1$ be the first order language containing no function letters, no individual constants, and only one predicate letter A_1.

Proposition 7.42

$K_{\mathscr{L}_1}$ is recursively decidable.

Proof. We describe an algorithm for answering questions from the set

$$\{\text{is } \mathscr{A} \text{ a theorem of } K_{\mathscr{L}_1}? \mid \mathscr{A} \text{ a } wf. \text{ of } K_{\mathscr{L}_1}\}.$$

To do this, we use the result from Chapter 4 that a *wf.* $\mathscr{A}$ is a theorem of $K_{\mathscr{L}_1}$ if and only if $\mathscr{A}$ is true in every interpretation of $\mathscr{L}_1$.

Let I be an interpretation of $\mathscr{L}_1$. Then $D_I = D_0 \cup D_1$, where

$$D_0 = \{x \in D_I : \bar{A}_1^1(x) \text{ holds}\},$$

and

$$D_1 = \{x \in D_I : \bar{A}_1^1(x) \text{ does not hold}\}.$$

(N.B. Either D_0 or D_1, but not both, may be empty.)

Now define an interpretation I^* as follows. The domain of I^* is $\{D_0, D_1\}$ if both D_0 and D_1 are non-empty, or $\{D_0\}$ if $D_1 = \varnothing$, or $\{D_1\}$ if $D_0 = \varnothing$. The interpretation of A_1^1 is $\hat{A}_1^1$, say, where $\hat{A}_1^1(D_0)$ holds and $\hat{A}_1^1(D_1)$ does not hold. For any *wf.* $\mathscr{A}$ of $\mathscr{L}_1$, if $\mathscr{A}$ is true in I^* then $\mathscr{A}$ is true in I. This is proved by induction on the number of connectives and quantifiers in $\mathscr{A}$, but note that different proofs are required for the different cases where I^* has domain $\{D_0\}$, $\{D_0, D_1\}$ or $\{D_1\}$.

It follows from the above that if a *wf.* $\mathcal{A}$ is true in every interpretation whose domain contains at most two elements, then $\mathcal{A}$ is true in all interpretations. Our algorithm, then, must be a method for checking whether *wfs.* are true in every interpretation with domain consisting of one or two elements. Every such interpretation must come into one of the following categories.

1.   Domain $\{d\}$, where $\bar{A}_1^1(d)$ holds.
2.   Domain $\{d\}$, where $\bar{A}_1^1(d)$ does not hold.
3.   Domain $\{d_1, d_2\}$, where $\bar{A}_1^1(d_1)$ holds, $\bar{A}_1^1(d_2)$ does not hold.
4.   Domain $\{d_1, d_2\}$, where $\bar{A}_1^1(d_1)$, $\bar{A}_1^1(d_2)$ both hold.
5.   Domain $\{d_1, d_2\}$, where $\bar{A}_1^1(d_1)$, $\bar{A}_1^1(d_2)$ both do not hold.

For any given *wf.* $\mathcal{A}$, it is easy to check whether $\mathcal{A}$ is true in interpretations in each category, using the methods of Chapter 3.

This algorithm may be used to decide whether any given *wf.* is a theorem, so $K_{\mathscr{L}_1}$ is recursively decidable.

$\triangleright$ The ideas of the above proof can be extended to give the much more general result:

Proposition 7.43

Let $\mathscr{L}$ be a first order language with no function letters or individual constants, and including only one-place predicate letters (possibly an infinite list of them). Then $K_{\mathscr{L}}$ is recursively decidable.

Proof. See Mendelson.

$\triangleright$ (A system such as $K_{\mathscr{L}}$ described above is often called a system of *pure* first order predicate calculus, indicating the absence of function letters and individual constants.)

In contrast to the above we now have the following proposition.

Proposition 7.44

The system $\mathcal{N}$ is recursively undecidable (under the assumption that it is consistent).

Proof. Let T be a unary relation on D_N defined by: $T(n)$ holds if and only if n is the Gödel number of a theorem of $\mathcal{N}$. Suppose that T is recursive (i.e. that $\mathcal{N}$ is recursively decidable). Then T is expressible in $\mathcal{N}$, by Proposition 6.12, and so there is a *wf.* $\mathcal{T}(x_1)$ in which x_1 occurs free, such that

$$\text{if } T(n) \text{ holds then } \underset{\mathcal{N}}{\vdash} \mathcal{T}(0^{(n)}),$$

and

> if $T(n)$ does not hold then $\vdash_{\mathcal{N}} \sim \mathcal{T}(0^{(n)})$.

Let D be the binary relation on D_N defined by: $D(m, n)$ holds if and only if either m is the Gödel number of a wf. $\mathcal{A}(x_1)$ in which x_1 occurs free, and n is the Gödel number of $\mathcal{A}(0^{(m)})$, or m is not the Gödel number of such a wf. and $n = 0$. Then D is a recursive function and is therefore representable in $\mathcal{N}$ (see Proposition 6.29), say by the wf. $\mathcal{D}(x_1, x_2)$.

Denote by $\mathcal{A}(x_1)$ the wf.

$$(\forall x_2)(\mathcal{D}(x_1, x_2) \to \sim \mathcal{T}(x_2)).$$

Let s be the Gödel number of this wf. Then $\mathcal{A}(0^{(s)})$ is the wf.

$$(\forall x_2)(\mathcal{D}(0^{(s)}, x_2) \to \sim \mathcal{T}(x_2)).$$

Let t be the Gödel number of this wf. Then $D(s, t)$ holds, and so $\vdash_{\mathcal{N}} \mathcal{D}(0^{(s)}, 0^{(t)})$. Now if $\mathcal{A}(0^{(s)})$ is a theorem of $\mathcal{N}$, then using axiom $(K5)$, we obtain

$$\vdash_{\mathcal{N}} (\mathcal{D}(0^{(s)}, 0^{(t)}) \to \sim \mathcal{T}(0^{(t)})).$$

It follows that

$$\vdash_{\mathcal{N}} \sim \mathcal{T}(0^{(t)}), \text{ by } MP.$$

Also, if $\mathcal{A}(0^{(s)})$ is not a theorem of $\mathcal{N}$, then t is not the Gödel number of a theorem of $\mathcal{N}$, so $T(t)$ does not hold, so $\vdash_{\mathcal{N}} \sim \mathcal{T}(0^{(t)})$. Thus, in any case, we have

$$\vdash_{\mathcal{N}} \sim \mathcal{T}(0^{(t)}).$$

Now

$$\vdash_{\mathcal{N}} \mathcal{D}(0^{(s)}, 0^{(t)}),$$

and since D is representable in $\mathcal{N}$, we have

$$\vdash_{\mathcal{N}} (\exists_1 x_2)\mathcal{D}(0^{(s)}, x_2).$$

Hence,

$$\underset{\mathcal{N}}{\vdash} (\mathscr{D}(0^{(s)}, x_2) \to x_2 = 0^{(t)}). \tag{*}$$

Our axioms for equality give

$$\underset{\mathcal{N}}{\vdash} (x_2 = 0^{(t)} \to (\sim \mathscr{T}(0^{(t)}) \to \sim \mathscr{T}(x_2))),$$

and since $\underset{\mathcal{N}}{\vdash} \sim \mathscr{T}(0^{(t)})$, we have

$$\underset{\mathcal{N}}{\vdash} (x_2 = 0^{(t)} \to \sim \mathscr{T}(x_2)).$$

With (*), using the rule *HS*, we obtain

$$\underset{\mathcal{N}}{\vdash} (\mathscr{D}(0^{(s)}, x_2) \to \sim \mathscr{T}(x_2)),$$

and by Generalisation,

$$\underset{\mathcal{N}}{\vdash} (\forall x_2)(\mathscr{D}(0^{(s)}, x_2) \to \sim \mathscr{T}(x_2)),$$

i.e.

$$\underset{\mathcal{N}}{\vdash} \mathscr{A}(0^{(s)}).$$

Thus t is the Gödel number of a theorem of $\mathcal{N}$, so

$$\underset{\mathcal{N}}{\vdash} \mathscr{T}(0^{(t)}).$$

But $\underset{\mathcal{N}}{\vdash} \sim \mathscr{T}(0^{(t)})$, so this contradicts the consistency of $\mathcal{N}$. The proof is therefore complete.

▷ If a system is known to be recursively decidable (or undecidable), can we deduce anything about the recursive decidability (or undecidability) of extensions of it? Recall that an extension of a first order system has the same language but a larger class of theorems. On the face of it, there would seem to be no reason why existence (or non-existence) of an algorithm which decides membership of the one class of theorems should imply existence (or non-existence) of an algorithm which decides membership of the other. (A recursive set can have a non-recursive subset, and a non-recursive set can have a recursive subset.) However, we can make a connection in certain circumstances.

Proposition 7.45

Let S and S^+ be first order systems with the same language, and let S^+ be a *finite* extension of S, i.e. suppose that there is a finite set $\mathscr{A}_1, \ldots, \mathscr{A}_n$ of *wfs.* such that when these are added to the axioms of S we obtain a set of axioms for S^+. If S^+ is recursively undecidable then S is also recursively undecidable.

Proof. Suppose that S and S^+ are as described, and that S^+ is recursively undecidable. Without loss of generality we may suppose that $\mathscr{A}_1, \ldots, \mathscr{A}_n$ are closed *wfs.* A proof in S^+ is a deduction in S from $\{\mathscr{A}_1, \ldots, \mathscr{A}_n\}$, and so for any *wf.* $\mathscr{A}$,

$$\underset{S^+}{\vdash} \mathscr{A} \text{ if and only if } \underset{S}{\vdash} (\mathscr{A}_1 \to (\ldots \to (\mathscr{A}_n \to \mathscr{A}))\ldots)$$

using the Deduction Theorem. If there were an algorithm for deciding whether any given *wf.* is a theorem of S then we could decide whether any given *wf.* $\mathscr{A}$ is a theorem of S^+ simply by asking whether $(\mathscr{A}_1 \to (\ldots \to (\mathscr{A}_n \to \mathscr{A}))\ldots)$ is a theorem of S. But S^+ is recursively undecidable, and so S must be recursively undecidable also.

▷ To summarise, then, we have discovered two distinct limitations of the system $\mathscr{N}$. It is not complete, and it is recursively undecidable. Firstly, the set of theorems of $\mathscr{N}$ does not coincide with the set of true statements, and secondly there is no algorithm for deciding which statements correspond to theorems of $\mathscr{N}$. So it would seem that $\mathscr{N}$ is not very helpful in providing a method for deciding the truth or falsity of statements in arithmetic. Formal systems of arithmetic, as discussed in this book, suffer from these limitations, and as yet there is no alternative approach.

Now for another major result (cf., Proposition 7.42).

Proposition 7.46

There is a first order language $\mathscr{L}$ such that $K_{\mathscr{L}}$ is recursively undecidable.

Proof. The proof uses ideas similar to those of the proof of Proposition 7.37. Let T be a Turing machine which halts if and only if the input number belongs to the set K (which is recursively enumerable but not recursive). Suppose that T uses the tape symbols 1 and B only, and that its internal states are $q_0, \ldots, q_n$. Now let $\mathscr{L}$ be a first order language whose alphabet of symbols includes as individual constants all the symbols in the set

$$A = \{B, \ 1, \ h, \ q, \ q', \ q_0, \ldots, q_n\},$$

and also includes the function letter f_2^2 and the predicate letter A_2^2. We shall think of f_2^2 as enabling us to form words, i.e. $f_2^2(x_1, x_2)$ is to be thought of as the word $x_1 x_2$. Then terms of $\mathcal{L}$ will correspond to words of any length (allowing occurrences of variables), and the set of closed terms of $\mathcal{L}$ involving only the symbols of A and the letter f_2^2 will correspond to the set S_A. (Any given word will correspond to several different terms, due to different possible bracketings, but we take account of this below.)

The predicate letter A_2^2 is to be thought of as the equivalence relation on S_A generated by the relations listed for the semigroup with unsolvable word problem described in the proof of Proposition 7.37.

The characteristics of f_2^2 and A_2^2 must be described, and the following *wfs.* of $K_{\mathcal{L}}$ do this.

(U1) $A_2^2(f_2^2(f_2^2(x_1, x_2), x_3), f_2^2(x_1, f_2^2(x_2, x_3)))$.

(U2) $A_2^2(x_1, x_2) \rightarrow A_2^2(f_2^2(x_1, x_3), f_2^2(x_2, x_3))$.

(U3) $A_2^2(x_1, x_2) \rightarrow A_2^2(f_2^2(x_3, x_1), f_2^2(x_3, x_2))$.

(U4)   For each $W = W'$ from the set of relations determining the semigroup with unsolvable word problem, we obtain closed terms t, t' of $K_{\mathcal{L}}$ corresponding to the words W and W', and include as (U4) all the *wfs.*

$$A_2^2(t, t')$$

obtained in this way. Note that there are only finitely many such.

(U5) $A_2^2(x_1, x_2) \rightarrow (A_2^2(x_2, x_3) \rightarrow A_2^2(x_1, x_3))$.

If we consider the first order system $K_{\mathcal{L}}^*$ obtained from $K_{\mathcal{L}}$ by including all the *wfs.* (U1)–(U5) as new axioms then it should be clear that for any closed terms t_1, t_2 of $\mathcal{L}$ corresponding to words W_1, W_2 of S_A, we have

$$\vdash_{K_{\mathcal{L}}^*} A_2^2(t_1, t_2) \quad \text{if and only if } W_1 \text{ and } W_2 \text{ are equivalent.}$$

In particular,

$$\vdash_{K_{\mathcal{L}}^*} A_2^2(t_1, f_2^2(h, f_2^2(q', h))) \quad \text{if and only if}$$

W_1 is equivalent to $h \ q' \ h$.

Now suppose that $K_{\mathscr{L}}^*$ is recursively decidable, i.e. there is an algorithm for deciding, for any wf. $\mathscr{A}$, whether $\vdash_{K_{\mathscr{L}}^*} \mathscr{A}$. Then given any word $W \in S_A$, to decide whether W is equivalent to $h\ q'\ h$ in S_A all we need do is form the wf. $A_2^2(t, f_2^2(h, f_2^2(q', h)))$ (where t corresponds to W) and ask whether this is a theorem of $K_{\mathscr{L}}^*$. Thus we have an algorithm for deciding, for any word W of S_A, whether it is equivalent to $h\ q'\ h$, and this contradicts our previous result. Hence $K_{\mathscr{L}}^*$ is recursively undecidable. But $K_{\mathscr{L}}^*$ is a finite extension of $K_{\mathscr{L}}$ (since $(U1)$–$(U5)$ constitute a finite set of wfs.). By Proposition 7.45, then, $K_{\mathscr{L}}$ is recursively undecidable.

▷ The reader should be careful to keep this result in the context of Proposition 7.42, and remember that whether a system of predicate calculus is recursively decidable or undecidable depends on the language $\mathscr{L}$. It should be noted, however, that undecidability is the rule rather than the exception, and indeed the above proof can be easily modified to show that if $\mathscr{L}$ contains at least one two-place function letter, at least one two-place predicate letter, and an infinite list of individual constants, then $K_{\mathscr{L}}$ is recursively undecidable.

Corollary 7.47

The full first order predicate calculus (with all of the symbols given in Chapter 3) is recursively undecidable.

▷ Let us end with some examples, which indicate that for systems with mathematical significance, recursive undecidability is most usual.

Proposition 7.48

(i) The following systems are recursively undecidable.
   (a) First order group theory.
   (b) First order theory of rings.
   (c) First order theory of fields.
   (d) First order theory of semigroups.
   (e) The system ZF.
(ii) The following systems are recursively decidable.
   (a) First order theory of *abelian* groups.
   (b) First order arithmetic without multiplication (i.e. the system which is the same as $\mathcal{N}$ except that the symbol f_2^2 is not included, and axioms $(N5)$ and $(N6)$ are omitted).

Proof. We do not go into the proofs, but the interested reader is referred to the book by Tarski, Mostowski and Robinson.

▷ Recursive decidability of a formal system implies the existence of a computer program which will decide, given any *wf.* of the system, whether it is a theorem or not (presuming that the computer is large enough, of course). Thus, for example, it is possible, using a machine and a single program, to decide whether statements about abelian groups and their elements are theorems or not. Such a program would be exceedingly complex and decisions about complicated statements would require considerable computer time and capacity, so this is not useful in a practical sense, but the possibility is of interest. So also is the impossibility of finding a corresponding program for the theory of groups and the other systems listed above as recursively undecidable. In particular, the recursive undecidability of *ZF* implies that there is no universal program which may be used to tell whether mathematical statements in general are theorems or not. Computers may run the world eventually, but they can never replace mathematicians!

Exercises

23 (See Proposition 7.42). Let $\mathscr{L}_2$ be the first order language with no individual constants, no function letters and only two one-place predicate letters A_1^1 and A_2^1. Show that a *wf.* $\mathscr{A}$ of $\mathscr{L}_2$ is true in every interpretation if it is true in every interpretation with domain containing four elements or fewer. Describe an algorithm for deciding whether a given *wf.* $\mathscr{A}$ of $\mathscr{L}_2$ is true in every such interpretation.

24 Prove that the following sets are not recursive.
   (*a*) $\{n \in D_N : n$ is the Gödel number of a *wf.* $\sim\!\mathscr{A}$, where $\mathscr{A}$ is a theorem of $\mathscr{N}\}$
   (*b*) $\{n \in D_n : n$ is the Gödel number of a *wf.* $\mathscr{A}$ of $\mathscr{N}$ which is false in $N\}$
   (*c*) $\{n \in D_N : n$ is the Gödel number of a *wf.* $\mathscr{A}$ of which is not a theorem of $\mathscr{N}\}$

25 Of the sets given in Exercise 24, show that (i) is recursively enumerable and that (iii) is not recursively enumerable.

26 Prove that the set (ii) in Exercise 24 is not recursively enumerable, as follows. Suppose the contrary, that the set is enumerated by the recursive function f. Define a relation F on D_N by: $F(m, n)$ holds if and only if m is the Gödel number of a *wf.* $\mathscr{A}(x_1)$ with one free variable x_1 and $f(n)$ is the Gödel number of $\mathscr{A}(0^{(m)})$. Show that F is recursive (using Church's Thesis). Then F is representable in $\mathscr{N}$ by the *wf.* $\mathscr{F}(x_1, x_2)$, say. Now let p be the Gödel number of the *wf.* $(\exists x_2)\mathscr{F}(x_1, x_2)$, and let $\mathscr{V}$ be the *wf.* $(\exists x_2)\mathscr{F}(0^{(p)}, x_2)$. Prove that $\mathscr{V}$ is false in N and that the Gödel number of $\mathscr{V}$ is not in the range of f, giving a contradiction. (Cf. the proof of the Gödel Incompleteness Theorem in Section 6.5.)

27 Let us say that a first order system S is *recursively axiomatisable* if there is a first order system T having the same theorems as S, such that the set of

Gödel numbers of axioms of T is recursive. Show that if S is recursively axiomatisable, then the set of Gödel numbers of theorems of S is recursively enumerable. Deduce that if S is recursively axiomatisable and complete, then S is recursively decidable.

Appendix. Countable and uncountable sets

Definition A1

A set is *countable* if it can be put in one–one correspondence with the set of natural numbers. In other words, a set A is countable if there is a bijection $f\colon D_N \to A$.

Notice that the elements of a countable set may be written out in a list, and that the bijection given in the definition provides a method for doing this, namely enumerating $f(0), f(1), f(2), \ldots$

Trivially, the natural numbers constitute a countable set.

Proposition A2

If A and B are countable sets then there is a bijection between A and B. Conversely, if A is a countable set and there is a bijection between A and B, then B is countable.

Proof. Let $f\colon D_N \to A$ and $g\colon D_N \to B$ be bijections. Then $g \circ f^{-1}$ is a bijection from A to B.

For the converse, let $f\colon D_N \to A$ be a bijection, so that A is countable, and suppose that there is a bijection h from A to B. Then $h \circ f$ is a bijection from D_N to B, and so B is countable.

Proposition A3

Any infinite subset of a countable set is countable.

Proof. Let A be a countable set, let $f\colon D_N \to A$ be a bijection and let B be an infinite subset of A. Then $f(0)$, $f(1)$, $f(2)$, $\ldots$ is a list of all the members of A. Delete from this list all elements which are not members of B. What remains is a list (infinite) of the members of B. A bijection $g\colon D_N \to B$ can now be defined by

$$g(n) = \text{the } (n+1)\text{th member of the new list } (n \in D_N).$$

(It must be the $(n+1)$th member because $g(0)$ is the first, $g(1)$ is the second, and so on.)

Proposition A4

An infinite set A is countable if and only if there is an injection $h: A \to D_N$.

Proof. If A is countable then there is a bijection $D_N \to A$, whose inverse is certainly an injection $A \to D_N$.

Conversely, suppose that there is an injection $h: A \to D_N$. $h(A) \subseteq D_N$ and $h(A)$ is infinite, since h is one–one. By Proposition A3, $h(A)$ is countable. Let $g: D_N \to h(A)$ be a bijection. The composition $h^{-1} \circ g$ is then a bijection from D_N to A, and A is countable.

▷ This last result is usually the most convenient to use in a demonstration that a particular set is countable, and we shall see applications of it shortly.

Proposition A5

The union of two disjoint countable sets is countable.

Proof. Let A and B be disjoint countable sets and let $f: D_N \to A$ and $g: D_N \to B$ be bijections. Define $h: D_N \to A \cup B$ as follows:

$$h(n) = \begin{cases} f(\tfrac{1}{2}n) & \text{if } n \text{ is even,} \\ g(\tfrac{1}{2}n - \tfrac{1}{2}) & \text{if } n \text{ is odd.} \end{cases}$$

It is easy to verify that h is a bijection, and so $A \cup B$ is countable. (h gives the list $f(0), g(0), f(1), g(1), f(2), \ldots$ of the elements of $A \cup B$.)

Corollary A6

The union of any finite collection of disjoint countable sets is countable.

Proof. The proof is by induction on the number of sets in the collection.

Base step: The union of two disjoint countable sets is countable, by the proposition.

Induction step: Let $n > 2$ and let $A_1, \ldots A_n$ be disjoint countable sets. Suppose as induction hypothesis that a union of $n - 1$ disjoint countable sets is countable. Then $A_1 \cup A_2 \cup \ldots \cup A_{n-1}$ is countable (and disjoint from A_n). By the proposition, then, the set $(A_1 \cup \ldots \cup A_{n-1}) \cup A_n$ is countable.

The result follows by the principle of mathematical induction.

Remark. The requirement that the sets be disjoint can be dispensed with. A demonstration of this is left as an exercise for the reader.

Question. Do there exist sets which are neither finite nor countable? We know from the above that D_N and all of its subsets are finite or countable. The answer is given in the following proposition.

Proposition A7

The set of all subsets of D_N is infinite and not countable.

Proof. Denote by $P(D_N)$ the set of all subsets of D_N. $P(D_N)$ is clearly infinite. Suppose that it is countable, and let $f: D_N \to P(D_N)$ be a bijection. Then for each $n \in D_N$, $f(n)$ is a subset of D_N. Let

$$B = \{k \in D_N : k \notin f(k)\}.$$

B is certainly a subset of D_N (possibly empty and possibly all of D_N). Also $B \neq f(n)$ for any $n \in D_N$. For suppose that $B = f(n)$. If $n \in f(n)$ then $n \in B$, since $B = f(n)$, but $n \notin B$ by the definition of B. If $n \notin f(n)$, then $n \notin B$, since $B = f(n)$, but $n \in B$ by definition of B. Either way we reach a contradiction. Therefore $B \neq f(n)$ for any $n \in D_N$, and so f is not a bijection between D_N and $P(D_N)$. This contradicts our original assumption, and hence $P(D_N)$ is not countable.

Corollary A8

The set of all functions on D_N is not countable.

Proof. For each subset A of D_N, define a function $C_A: D_N \to D_N$ (the characteristic function of A) as follows.

$$C_A(n) = \begin{cases} 0 & \text{if } n \in A, \\ 1 & \text{if } n \notin A. \end{cases}$$

The correspondence between sets A and functions C_A is a bijection from $P(D_N)$ to a *subset* of the set of functions on D_N. $P(D_N)$ is not countable, so the set of functions on D_N has a subset which is not countable. (If there is a bijection between two sets and one of them is countable, then the other is countable.) If the set of functions on D_N were countable, then any infinite subset would be countable (by Proposition A3). Hence, since it has an uncountable subset, the set of all functions on D_N is not countable.

Corollary A9

The set of all relations on D_N is not countable.

*Proof.*The set of relations includes the set of functions. By an argument similar to the above, then, the set of relations cannot be countable.

▷ In the text we use the result that if an uncountable set has a countable subset, then that subset must be a proper subset. This should now be clear, since a set cannot be both countable and uncountable.

We also use the result that the set of *wfs.* in a given symbolic language was countable. There are some general results which lead us to see why this is so.

Proposition A10

Let A be a countable set. The collection of all *finite* subsets of A is a countable set.

Proof. Let $f: D_N \to A$ be a bijection. We may define an injection g from the set of all finite subsets of A into D_N as follows. Let F be a finite subset of A. Then $f^{-1}(F)$ is a finite subset of D_N. Let $g(F) =$ the product of the primes p_n, for $n \in f^{-1}(F)$. (Here p_i denotes the ith odd prime, for $i > 0$, and $p_0 = 2$.) g is one–one since the same product of primes cannot possibly arise from two different sets F, and two different products of primes cannot be equal. By Proposition A4, then, the set of all finite subsets of D_N is countable.

Proposition A11

Let A be a countable set. Then the set of all finite *sequences of* elements of A is a countable set.

Proof. We can make use of the properties of prime numbers here in a slightly different way. Let $f: D_N \to A$ be a bijection. Define an injection h from the set of all finite sequences of elements of A into D_N as follows. If $u_0, u_1, \ldots, u_k \in A$, let

$$h(u_0, u_1, \ldots, u_k) = p_0^{f^{-1}(u_0)} \times p_1^{f^{-1}(u_1)} \times \ldots \times p_k^{f^{-1}(u_k)},$$

where the p_i are as in the previous proof. h is one–one because f is a bijection and because of the uniqueness of prime power decomposition. By Proposition A4, then, the set of all finite sequences of elements of A is countable.

▷ This proposition has as a corollary the result which we need about formal languages. All of our formal languages have alphabets of symbols which are countable sets. (Demonstration of this requires use of Corollary A6.) The set of *wfs.* of a formal language $\mathcal{L}$ is a subset of the set of all finite sequences of symbols from the alphabet of $\mathcal{L}$. This set of finite sequences is countable, so any subset of it is also countable (or finite). But we know that the set of all *wfs.* is always infinite, and so we have our result.

Hints and solutions to selected exercises

Chapter 1

Section 1.1 (p. 3)

1(*a*) $(A \wedge B) \rightarrow C$.
  (*g*) $A \leftrightarrow (B \vee C)$.
  (*h*) $A \rightarrow (B \rightarrow C)$.

Section 1.2 (p. 10)

5   (*a*), (*b*)
7   Take *p* true and *q* true. Take $\mathcal{A}$ and $\mathcal{B}$ to be any tautologies.

Section 1.3 (p. 15)

10        $((\sim p) \vee q)$ is logically equivalent to $(p \rightarrow q)$,
        so $(\sim((\sim p) \vee q)) \vee r)$ is logically equivalent to $(\sim(p \rightarrow q) \vee r)$.

Section 1.4 (pp. 18–19)

12(*a*) $((p \wedge q) \vee ((\sim p) \wedge (\sim q)))$.
  (*d*) $((p \wedge (\sim q) \wedge (\sim r)) \vee ((\sim p) \wedge q \wedge (\sim r)) \vee ((\sim p) \wedge (\sim q) \wedge (\sim r)))$.
13(*a*) $((((\sim p) \vee (\sim q) \vee r) \wedge (p \vee (\sim q) \vee r) \wedge (p \vee q \vee r))$.

Section 1.5 (pp. 21–2)

14(*b*) $((p \vee q) \vee (\sim (r \vee (\sim s))))$.
17(*b*) Consider a statement form involving statement variables *p* and *q* and the connectives $\sim$ and $\leftrightarrow$ only. Show that its truth table must contain four Fs or four Ts or two Fs and two Ts. A full demonstration requires a proof by induction of the sort used in Proposition 1.15.

19 Such a truth table must have four rows, and so must be one of sixteen possible. All sixteen can be accounted for as known or not adequate.

Section 1.6 (pp. 25–6)

20(*b*) $(p \rightarrow (q \lor r))$, $(\sim r)$; $\therefore$ $((\sim q) \rightarrow (\sim p))$.   Valid.

(*d*) $(p \rightarrow (q \land r \land s))$, q, $(s \rightarrow r)$; $\therefore$ $s \rightarrow p$.   Invalid.

Chapter 2

Section 2.1 (pp. 36–7)

1(*b*) (1) $(p_1 \rightarrow (p_2 \rightarrow p_3)) \rightarrow ((p_1 \rightarrow p_2) \rightarrow (p_1 \rightarrow p_3))$ (*L*2)

(2) $((p_1 \rightarrow (p_2 \rightarrow p_3)) \rightarrow ((p_1 \rightarrow p_2) \rightarrow (p_1 \rightarrow p_3))) \rightarrow (((p_1 \rightarrow (p_2 \rightarrow p_3))$
$\rightarrow (p_1 \rightarrow p_2)) \rightarrow ((p_1 \rightarrow (p_2 \rightarrow p_3)) \rightarrow (p_1 \rightarrow p_3)))$ (*L*2)

(3) $((p_1 \rightarrow (p_2 \rightarrow p_3)) \rightarrow (p_1 \rightarrow p_2)) \rightarrow ((p_1 \rightarrow (p_2 \rightarrow p_3))$
$\rightarrow (p_1 \rightarrow p_3))$ 1, 2 *MP*

(*d*) (1) $p_2 \rightarrow (p_1 \rightarrow p_2)$ (*L*1)

(2) $(p_2 \rightarrow (p_1 \rightarrow p_2)) \rightarrow (p_1 \rightarrow (p_2 \rightarrow (p_1 \rightarrow p_2)))$ (*L*1)

(3) $(p_1 \rightarrow (p_2 \rightarrow (p_1 \rightarrow p_2)))$ 1, 2 *MP*

2(*b*) (1) $(\sim \sim \mathscr{A})$ assumption

(2) $(\sim \sim \mathscr{A}) \rightarrow ((\sim \sim \sim \sim \mathscr{A}) \rightarrow (\sim \sim \mathscr{A}))$ (*L*1)

(3) $(\sim \sim \sim \sim \mathscr{A}) \rightarrow (\sim \sim \mathscr{A})$ 1, 2 *MP*

(4) $((\sim \sim \sim \sim \mathscr{A}) \rightarrow (\sim \sim \mathscr{A})) \rightarrow ((\sim \mathscr{A}) \rightarrow (\sim \sim \sim \mathscr{A}))$ (*L*3)

(5) $(\sim \mathscr{A}) \rightarrow (\sim \sim \sim \mathscr{A})$ 3, 4 *MP*

(6) $((\sim \mathscr{A}) \rightarrow (\sim \sim \sim \mathscr{A})) \rightarrow ((\sim \sim \mathscr{A}) \rightarrow \mathscr{A})$ (*L*3)

(7) $(\sim \sim \mathscr{A}) \rightarrow \mathscr{A}$ 5, 6 *MP*

(8) $\mathscr{A}$ 1, 7 *MP*

(*d*) (1) $\mathscr{A} \rightarrow (\mathscr{B} \rightarrow \mathscr{C})$ assumption

(2) $(\mathscr{A} \rightarrow (\mathscr{B} \rightarrow \mathscr{C})) \rightarrow ((\mathscr{A} \rightarrow \mathscr{B}) \rightarrow (\mathscr{A} \rightarrow \mathscr{C}))$ (*L*2)

(3) $(\mathscr{A} \rightarrow \mathscr{B}) \rightarrow (\mathscr{A} \rightarrow \mathscr{C})$ 1, 2 *MP*

(4) $\mathscr{B} \rightarrow (\mathscr{A} \rightarrow \mathscr{B})$ (*L*1)

(5) $\mathscr{B} \rightarrow (\mathscr{A} \rightarrow \mathscr{C})$ 3, 4 *HS*

3(*b*) (1) $\mathscr{B} \rightarrow \mathscr{A}$ assumption

(2) $(\sim \sim \mathscr{B}) \rightarrow \mathscr{B}$ by Exercise 2(*b*) and Ded. Thm.

(3) $(\sim \sim \mathscr{B}) \rightarrow \mathscr{A}$ 1, 2 *HS*

(4) $\mathscr{A} \rightarrow (\sim \sim \mathscr{A})$ by Exercise 3(*a*)

(5) $(\sim \sim \mathscr{B}) \rightarrow (\sim \sim \mathscr{A})$ 3, 4 *HS*

(6) $((\sim \sim \mathscr{B}) \rightarrow (\sim \sim \mathscr{A})) \rightarrow (\sim \mathscr{A} \rightarrow \sim \mathscr{B})$ (*L*3)

(7) $(\sim \mathscr{A}) \rightarrow (\sim \mathscr{B})$ 5, 6 *MP*

$\therefore$ $(\mathscr{B} \rightarrow \mathscr{A}) \underset{L}{\vdash} ((\sim \mathscr{A}) \rightarrow (\sim \mathscr{B}))$

so $\underset{L}{\vdash} (\mathscr{B} \rightarrow \mathscr{A}) \rightarrow ((\sim \mathscr{A}) \rightarrow (\sim \mathscr{B}))$ by the Ded. Thm.

3(*d*) (1) $\sim (\mathscr{A} \rightarrow \mathscr{B})$ assumption

(2) $(\mathscr{B} \rightarrow (\mathscr{A} \rightarrow \mathscr{B})) \rightarrow (\sim (\mathscr{A} \rightarrow \mathscr{B}) \rightarrow \sim \mathscr{B})$ by Exercise 3(*b*)

(3) $\mathcal{B} \to (\mathcal{A} \to \mathcal{B})$ (L1)
(4) $\sim(\mathcal{A} \to \mathcal{B}) \to \sim\mathcal{B}$ 2, 3 MP
(5) $\sim\mathcal{B}$ 1, 5 MP
(6) $\sim\mathcal{B} \to (\mathcal{B} \to \mathcal{A})$ Proposition 2.11(a)
(7) $(\mathcal{B} \to \mathcal{A})$ 5, 6 MP
∴ $\sim(\mathcal{A} \to \mathcal{B}) \underset{L}{\vdash} (\mathcal{B} \to \mathcal{A})$
so $\underset{L}{\vdash} (\sim(\mathcal{A} \to \mathcal{B}) \to (\mathcal{B} \to \mathcal{A}))$ by the Ded. Thm.

4   Note that the Deduction Theorem holds for L', since $(L3)$ is not used in its proof.
(i) (1) $(\sim\mathcal{A}) \to (\sim\mathcal{B})$ assumption
   (2) $(\sim\mathcal{A}) \to \mathcal{B}$ assumption
   (3) $((\sim\mathcal{A}) \to (\sim\mathcal{B})) \to (\mathcal{B} \to \mathcal{A})$ (L3)
   (4) $(\mathcal{B} \to \mathcal{A})$ 1, 3 MP
   (5) $(\sim\mathcal{A}) \to \mathcal{A}$ 2, 4 HS
   (6) $((\sim\mathcal{A}) \to \mathcal{A}) \to \mathcal{A}$ Proposition 2.11(b)
   (7) $\mathcal{A}$ 5, 6 MP
(ii) (1) $(\sim\mathcal{A}) \to (\sim\mathcal{B})$ assumption
    (2) $\mathcal{B}$ assumption
    (3) $((\sim\mathcal{A}) \to (\sim\mathcal{B})) \to (((\sim\mathcal{A}) \to \mathcal{B}) \to \mathcal{A})$ (L'3)
    (4) $((\sim\mathcal{A}) \to \mathcal{B}) \to \mathcal{A}$ 1, 3 MP
    (5) $B \to ((\sim\mathcal{A}) \to \mathcal{B})$ (L1)
    (6) $(\sim\mathcal{A}) \to \mathcal{B}$ 2, 5 MP
    (7) $\mathcal{A}$ 4, 6 MP

Section 2.2 (p. 44)

8   $\mathcal{A}$ is not a tautology, so is not a theorem of L. Apply Exercise 7. L^+ is consistent, for suppose otherwise. Then $\underset{L}{\vdash} (\mathcal{A} \to \mathcal{B})$ and $\underset{L}{\vdash} (\mathcal{A} \to \sim\mathcal{B})$, for some wf. $\mathcal{B}$. Then $\underset{L}{\vdash} ((\sim\mathcal{B}) \to (\sim\mathcal{A}))$, and $\underset{L}{\vdash} (\mathcal{B} \to (\sim\mathcal{A}))$, and it follows that $\underset{L}{\vdash} (\sim\mathcal{A})$. But $\mathcal{A}$ is not a contradiction, so $(\sim\mathcal{A})$ is not a theorem of L.

10 Let $\mathcal{A}$ and $\mathcal{B}$ be tautologies. Then $((\sim\mathcal{A} \to \mathcal{B}) \to (\mathcal{A} \to \sim\mathcal{B}))$ is a contradiction, and its negation is a theorem of L, and so of L^{++}. But it is an instance of the new axiom scheme, so is itself a theorem of L^{++}. Thus L^{++} is inconsistent.

12 Use Propositions 1.10, 2.14 and 2.23.

Chapter 3

Section 3.1 (p. 49)

1(a) $\sim(\forall x)(F(x) \to D(x))$.
 (c) $(\exists x)(T(x) \land L(x)) \to (\forall x)(T(x) \to L(x))$.

(f) $(\exists x)(P(x) \wedge (\forall y)(P(y) \to H(x, y)))$.

2(a) $\sim(\forall x)(C(x) \to T(x))$
             $(\exists x)(C(x) \wedge \sim T(x))$.

(c) $(\forall x)(\forall y)((M(x) \wedge E(y)) \to \sim H(x, y))$
             $\sim(\exists x)(\exists y)(M(x) \wedge E(y) \wedge H(x, y))$.

Section 3.2 (p. 56)

6   (a), (e), (g), (h).

7   $f_1^2(x_1, x_2)$ is free for x_2 in all of these.

8   x_j occurs free in $\mathscr{A}(x_j)$ only where x_i occurred free in $\mathscr{A}(x_i)$. No such occurrence can be within the scope of a $(\forall x_i)$.

9   t is free for x_1 in (a) and (b).

10 (a) x_2 is free for x_1 in (b) and (c).
    (d) $f_1^3(x_1, x_2, x_3)$ is free for x_1 in (b) only.

Section 3.3 (p. 59)

11 Interpretation of $\mathscr{A}$ in I is true.
    Take D_I to be $\mathbf{Z}$, $\bar{a}_1$ to be 0, $\bar{f}_1^2$ to be $+$, $\bar{A}_2^2$ to be $=$.

12 Take D_I to be $\mathbf{Z}$, $\bar{A}_1^1(x)$ to be $x > 0$, $\bar{f}_1^1(x)$ to be $-x$.

Section 3.4 (pp. 69–70)

14(a) v satisfies if $v(x_1) = 4$, $v(x_2) = 2$, $v(x_3) = 6$;
            v does not satisfy if $v(x_1) = 1$, $v(x_2) = 2$, $v(x_3) = 4$.

(b) v satisfies if $v(x_1) = 1$, $v(x_2) = 1$, $v(x_3) = 2$;
            v does not satisfy if $v(x_1) = 1$, $v(x_2) = 1$, $v(x_3) = 3$.

15(b) v satisfies if $v(x_1) = 1$, $v(x_2) = 2$;
            v does not satisfy if $v(x_1) = 3$, $v(x_2) = 2$.

(c) v satisfies if $v(x_1) = 1$, $v(x_2) = 1$;
            v does not satisfy if $v(x_1) = 1$, $v(x_2) = 2$.

16 (a) false; (b), (c), (d) true.

19(a) Let I be an interpretation and let v be a valuation in I satisfying $(\exists x_1)(\forall x_2)A_1^2(x_1, x_2)$. Then there is v', 1-equivalent to v, which satisfies $(\forall x_2)A_1^2(x_1, x_2)$, say with $v'(x_1) = x$. Then every valuation v'' which is 2-equivalent to v' satisfies $A_1^2(x_1, x_2)$. Now suppose that v does not satisfy $(\forall x_2)(\exists x_1)A_1^2(x_1, x_2)$. Then there is a valuation w, 2-equivalent to v, which does not satisfy $(\exists x_1)A_1^2(x_1, x_2)$ with (say) $w(x_2) = y$. There is no valuation 1-equivalent to w which satisfies $A_1^2(x_1, x_2)$. But if $v''(x_1) = x$, $v''(x_2) = y$, $v''(x_k) = v(x_k)$ $(k > 2)$, then v'' is 1-equivalent to w and satisfies $A_1^2(x_1, x_2)$.

(c) Let I be an interpretation and let v be a valuation in I satisfying $(\forall x_1)(\mathscr{A} \to \mathscr{B})$. Then every v', 1-equivalent to v, satisfies $(\mathscr{A} \to \mathscr{B})$. Now suppose that v does not satisfy $(\forall x_1)\mathscr{A} \to (\forall x_1)\mathscr{B}$. Then v satisfies $(\forall x_1)\mathscr{A}$ and v does not satisfy $(\forall x_1)\mathscr{B}$, and so there is a v', 1-equivalent to v, which satisfies $\mathscr{A}$ and does not satisfy $\mathscr{B}$. This v' does not satisfy $\mathscr{A} \to \mathscr{B}$.

21 Use Proposition 3.23.

22(a) Take D_I to be $\mathbf{Z}$, $\bar{A}_1^2$ to be $<$.

  (c) Take D_I to be $\mathbf{Z}$, $\bar{A}_1^2(x)$ to be $x = 0$, $\bar{a}_1$ to be 0.

  (d) This is false in N.

23 Proposition 3.23.

24(a) $(\forall x_1)(\forall x_3)\, A_1^3(x_1, h_1^1(x_1), x_3)$.

  (b) $(\forall x_2)\, A_1^2(c_1, x_2) \to A_2^2(c_1, c_2)$.

  (c) $(\forall x_1)(\sim A_1^1(x_1) \to \sim A_1^2(h_1^1(x_1), h_2^1(x_1)))$.

  (d) $(\forall x_1)(\forall x_3)((\sim A_1^2(x_1, h_1^1(x_1)) \vee A_2^1(h_1^1(x_1))) \to A_2^2(x_3, h_1^2(x_1, x_3)))$.

Chapter 4

Section 4.1 (p. 80)

1 See example 2.7(a). Use Generalisation once.

2(a) First show that $\{\sim \mathcal{B}, (\forall x_i)\mathcal{A}\} \underset{K}{\vdash} (\forall x_i) \sim (\mathcal{A} \to \mathcal{B})$. By the Deduction Theorem, deduce that

$$\sim \mathcal{B} \underset{K}{\vdash} ((\forall x_i)\mathcal{A} \to (\forall x_i) \sim (\mathcal{A} \to \mathcal{B}))$$

and hence that

$$\sim \mathcal{B} \underset{K}{\vdash} (\sim (\forall x_i) \sim (\mathcal{A} \to \mathcal{B}) \to \sim (\forall x_i)\mathcal{A})$$

Now by the converse to the Deduction Theorem,

$$\{\sim \mathcal{B}, (\exists x_i)(\mathcal{A} \to \mathcal{B})\} \underset{K}{\vdash} \sim (\forall x_i)\mathcal{A}$$

and we get $(\exists x_i)(\mathcal{A} \to \mathcal{B}) \underset{K}{\vdash} (\sim \mathcal{B} \to \sim (\forall x_i)\mathcal{A})$, provided x_i does not occur free in $\mathcal{B}$. Two more steps give the required result.

  (b) Similarly, first show that $\{((\exists x_i)\mathcal{A} \to \mathcal{B}), \sim \mathcal{B}\} \underset{K}{\vdash} \sim \mathcal{A}$. Deduce that $((\exists x_i)\mathcal{A} \to \mathcal{B}) \underset{K}{\vdash} (\mathcal{A} \to \mathcal{B})$, and apply Generalisation.

  (c) Sufficient to show $\underset{K}{\vdash} (((\forall x_i) \sim \sim \mathcal{A}) \to (\forall x_i)\mathcal{A})$.

3(b) Take D_I to be $\mathbf{Z}$, $\bar{A}_1^2$ to be $<$.

Section 4.2 (pp. 85–6)

4 Start: (1) $(\forall x_i)(\mathcal{A} \to \mathcal{B})$ assumption

        (2) $(\forall x_i)\mathcal{A}$ assumption.

5 Required to show that

$$\underset{K}{\vdash} (\sim (\sim (\forall x_i) \sim \mathcal{A}) \to (\forall x_i)(\sim \mathcal{A}))$$

and

$$\underset{K}{\vdash} ((\forall x_i)(\sim \mathcal{A}) \to (\sim (\forall x_i) \sim \mathcal{A})).$$

$6(a)$　　Use $(K5)$ twice and then Generalisation twice.

(b)　　Use $(K5)$ in the form $(\forall x_2)A_1^2(x_1, x_2) \to A_1^2(x_1, x_1)$.

7　First apply Proposition 4.18 to the *wf.* $\sim \mathcal{A}(x_i)$ to obtain

$$\vdash_K (\forall x_i) \sim \mathcal{A}(x_i) \leftrightarrow (\forall x_j) \sim \mathcal{A}(x_j).$$

Section 4.3 (p. 92)

$8(a)$　　$(\exists x_3)(\forall x_2)(A_1^1(x_3) \to A_1^2(x_1, x_2))$.

(c)　　$(\exists x_1)(\forall x_4)(\exists x_3)((A_1^1(x_1) \to A_1^2(x_1, x_2)) \to (A_1^1(x_4) \to A_1^2(x_4, x_3)))$.

　　(N.B. These answers are not unique.)

10　$((\exists x_1)(\exists x_2)A_1^2(x_1, x_2) \to (\exists x_3)A_1^1(x_3))$ is provably equivalent to *wfs.* in prenex form of both Π_3 form and Σ_2 form.

11　(a)　$\sim A_1^1(c_1) \lor A_1^2(x_1, x_2)$.

　　(c)　$(\sim A_1^1(c_1) \lor \sim A_1^2(c_1, x_2) \lor \sim A_1^1(x_4) \lor A_1^2(x_4, h_1^1(x_4)))$

　　　　$\land \; (A_1^1(c_1) \lor \sim A_1^2(c_1, x_2) \lor \sim A_1^1(x_4) \lor A_1^2(x_4, h_1^1(x_4)))$

　　　　$\land \; (A_1^1(c_1) \lor A_1^2(c_1, x_2) \lor \sim A_1^1(x_4) \lor A_1^2(x_4, h_1^1(x_4)))$.

Section 4.4 (p. 100)

12　See Proposition 2.18. The proof is essentially the same.

13　Assume S not complete and use Proposition 4.37 twice.

14　Take $\mathcal{B}$ to be $(\sim \mathcal{A})$. $K_{\mathcal{L}}$ is not complete, so we may take both $\mathcal{A}$ and $(\sim \mathcal{A})$ to be non-theorems.

15　No atomic formula or its negation can be a theorem of $K_{\mathcal{L}}$. By Proposition 4.37, then, including an atomic formula or its negation as a new axiom will give a consistent extension. Different predicate letters give different atomic formulas which give different extensions.

Section 4.5 (pp. 103–4)

16　Use induction on the number of steps in a deduction of $\mathcal{A}$ from Γ. Axioms of $K_{\mathcal{L}}$ and members of Γ are all true in M, and the rules of deduction preserve truth in M. (See the proof of Proposition 4.43.) The converse need not hold unless the extension of $K_{\mathcal{L}}$ obtained by including all the *wfs.* in Γ as new axioms is complete.

18　S^+ need not be complete. Take S to be $K_{\mathcal{L}}$, where $\mathcal{L}$ contains only one predicate letter A_1^1. M can be constructed so that *no* atomic formula is true in M, so $S^+ = S = K_{\mathcal{L}}$. Neither $(\forall x_1)A_1^1(x_1)$ nor $\sim (\forall x_1)A_1^1(x_1)$ is a theorem of $K_{\mathcal{L}}$.

19　All the new axioms are true in M, so by Exercise 16, every theorem of $\hat{S}$ is true in M. Hence S is consistent. S need not be complete.

20　See the proof of Proposition 4.48 (consider the additional axioms $A_1^1(a_i)$, $i \geq 1$).

Chapter 5

Section 5.2 (pp. 111–12)

1   Use Proposition 3.27 and the logical validity of axiom $(K5)$ to deduce that for any wf. $\mathscr{A}(x_i)$, if $I \models \mathscr{A}(x_i)$. then $I \models \mathscr{A}(t)$ for any term t which is free for x_i in $\mathscr{A}(x_i)$. Apply this to the wf. $A_1^2(x_1, x_2) \to A_1^2(f_1^2(x_1, x_3), f_1^2(x_2, x_3))$ (three times, to substitute terms for each of the variables) and similarly to $A_1^2(x_1, x_2) \to A_1^2(f_1^2(x_3, x_1), f_1^2(x_3, x_2))$.

2   Suppose the contrary, and use Propositions 4.37 and 5.6.

3   Induction on the length of $\mathscr{A}$. Base step is direct from $(E9')$. Induction step: Note that if

$$\vdash x_1 = x_2 \to (\mathscr{B}(x_1) \to \mathscr{B}(x_2))$$

then

$$\vdash x_1 = x_2 \to (\mathscr{B}(x_2) \to \mathscr{B}(x_1))$$

since

$$\vdash x_2 = x_1 \to (\mathscr{B}(x_2) \to \mathscr{B}(x_1))$$

and

$$\vdash (x_1 = x_2 \to x_2 = x_1)).$$

Use the Deduction Theorem.

4   From $A_1^2(u_j, v_j)$ $(1 \le j \le n)$ we can deduce $A_i^n(u_1, \dots, u_n) \leftrightarrow A_i^n(v_1, \dots, v_n)$ in S, by repeated use of $(E9)$, where u_j and v_j stand for variables. Hence if $\bar{A}_1^2(y_j, z_j)$ holds in M for each j, $\bar{A}_i^n(y_1, \dots, y_n)$ holds in M if and only if $\bar{A}_i^n(z_1, \dots, z_n)$ holds.

6   Inductive argument similar to Exercise 3.

Section 5.3 (p. 116)

7(a)   Language: variables, $f_{1}^1, f_1^2, =$.
Replace $(G2)$ and $(G3)$ by:

$$(\exists x_1)(\forall x_2)(f_1^2(x_1, x_2) = x_2 \wedge f_1^2(f_1^1(x_2), x_2) = x_1).$$

(b)   Language: variables, $a_1, =, A_1^3$.
($A_1^3(x_1, x_2, x_3)$ to be interpreted $x_1 x_2 = x_3$.)
Replace $(G1)$–$(G3)$ by

$$(\forall x_1)(\forall x_2)(\exists x_3) A_1^3(x_1, x_2, x_3),$$

$$(A_1^3(x_1, x_2, x_4) \wedge A_1^3(x_4, x_3, x_5) \wedge A_1^3(x_2, x_3, x_6)$$

$$\wedge A_1^3(x_1, x_6, x_7)) \to x_5 = x_7,$$

$$(\forall x_2)(\exists x_1) A_1^3(x_1, x_2, a_1)$$

and

$$(\forall x_1)A_1^3(a_1, x_1, x_1).$$

9 Including a_1 has the effect of distinguishing one element of the model, namely $\bar{a}_1$, and there is no significance to which element is chosen. Likewise with a sequence $a_1, a_2, \ldots$, and the interpretations of these need not all be different elements.

12 Denote by kx_1 the sum of x_1 with itself k times $(k > 1)$ in $\mathscr{F}$, and by (Ck) the wf. $(kx_1 = 0) \rightarrow (x_1 = 0)$ of $\mathscr{F}$. Each (Ck) is true in fields of characteristic zero, indeed $\mathscr{F}$ with these as additional axioms gives a system of the theory of fields of characteristic zero. Now if $\mathscr{A}$ is true in every model of this system, it is a theorem of the system. In a proof of $\mathscr{A}$, only finitely many axioms are used, so (say) no (Ck) with $k > n$ is used. Fields of characteristic p, with $p > n$ are models of $\mathscr{F}$ with the additional axioms $(C2), \ldots, (Cn)$. $\mathscr{A}$ is a theorem of this system, so is true in each such model.

13 Use ideas of Exercise 12, and see Corollary 4.49.

Section 5.4 (p. 120)

14 The axioms for $\mathscr{N}'$ do not mention a_0 specifically, so its interpretation is not constrained to have any particular properties. If $\bar{a}_0$ is chosen to be r, then the wf. $\sim(a_0 = a_i)$ is true in $\mathscr{N}'$ for every $i \leq r$. Thus for each r, the set of wfs. $\{\sim(a_0 = a_i): 0 < i \leq r\}$ has a model, and the system obtained from $\mathscr{N}'$ by including them as axioms is consistent. Hence the system obtained by including all these wfs. for $i > 0$ is consistent (by a familiar argument – see for example the proof of Proposition 2.21). Such a model is called a non-standard model of arithmetic, for it includes all the natural numbers and at least one other element. This other element must of course have successors and predecessors and sums and products with natural numbers and with itself, and so on.

Section 5.5 (p. 125)

15 Replace $(ZF2)$ by $(\forall x_1) \sim (x_1 \in a_1)$.
Add $(ZF2')$: $A_3^2(t_1, t_2) \leftrightarrow (\forall x_1)(x_1 \in t_1 \rightarrow x_1 \in t_2)$.
Replace $(ZF3)$ by $(\forall x_1)(\forall x_2)(\forall x_3)(x_3 \in f_1^2(x_1, x_2) \leftrightarrow x_3 = x_1 \vee x_3 = x_2)$.
The elements of the domain D of this interpretation are $0, 1, 2, \ldots$, where $0 = \emptyset$ and $n = \{0, 1, \ldots, n-1\}$ for $n \neq 0$. For $(ZF1)$: if $m = n = 0$ then both m and n have no members, if $m = n \neq 0$ then each has $0, 1, \ldots, m-1$ as members, and conversely, if m and n have the same members then either both are 0 or $\{0, 1, \ldots, m-1\} = \{0, 1, \ldots, n-1\}$, so $m = n$. For $(ZF2)$, clearly 0 is the null set. For $(ZF3)$, consider $2, 3 \in D$. $\{2, 3\} \notin D$, so $(ZF3)$ is false. For $(ZF4)$: $\bigcup 0 = 0$, $\bigcup 1 = 0$ and $\bigcup\{0, 1, \ldots, m-1\} = \{0, 1, \ldots, m-2\} = m - 1 \in D$. For $(ZF5)$: the subsets of the set 2 are $\emptyset, \{0\}, \{1\}$ and $\{0, 1\}$, and these elements do not constitute a member of D. For $(ZF7)$: suppose that m is the set whose existence is asserted by $(ZF7)$ – then $m \neq 0$ since $0 \notin 0$ and if $m = \{0, 1, \ldots, m-1\}$ then $m - 1 \in m$ and

$(m-1)\cup\{m-1\}\notin m$ because $m\notin m$. For $(ZF8)$, if $m=\{0,1,\ldots,m-1\}$, then $0\in m$ and 0 has no elements in common with m.

$(ZF6)$ is false. Take $\mathscr{A}(x_1,x_2)$ to be $x_2=\{x_1\}$. Then the image of the set 2 (for example) is not a member of D.

Chapter 6

Section 6.2 (pp. 136–7)

1 By Example 6.7, if $m+r=n$ then $\vdash_{\mathscr{N}} 0^{(m)}+0^{(r)}=0^{(n)}$. Using the axiom $(K5)$ in the form

$$(\forall x_1)\sim(0^{(m)}+x_1=0^{(n)})\to\sim(0^{(m)}+0^{(r)}=0^{(n)})$$

we can deduce $\vdash_{\mathscr{N}}\sim(\forall x_1)\sim(0^{(m)}+x_1=0^{(n)})$ as required.

2 Let $m>n$. Then $m=n+r$, with $r>0$. So $\vdash_{\mathscr{N}}0^{(m)}+0^{(n)}+0^{(r)}$ and $\vdash_{\mathscr{N}}0^{(m)}+x_1$
$=0^{(n)}+0^{(r)}+x_1$. Now $\vdash_{\mathscr{N}}0^{(n)}+0^{(r)}+x_1=0^{(n)}$ yields $\vdash_{\mathscr{N}}0^{(r)}+x_1=0$ (see the
proof of Proposition 6.1), and $\vdash_{\mathscr{N}}0^{(r)}+x_1=(0^{(r-1)}+x_1)'$ by $(N4^*)$, since
$r>0$. Hence

$$\vdash_{\mathscr{N}}0^{(n)}+0^{(r)}+x_1=0^{(n)}\to\sim(N1^*),$$

and so

$$\vdash_{\mathscr{N}}\sim(0^{(n)}+0^{(r)}+x_1=0^{(n)}),\text{ i.e. }\vdash_{\mathscr{N}}\sim(0^{(m)}+x_1=0^{(n)}).$$

By Generalisation, $\vdash_{\mathscr{N}}(\forall x_1)\sim(0^{(m)}+x_1=0^{(n)})$ and the result follows.

3(a) $\sim(\forall x_1)(\forall x_2)\sim(x_1\,x_2=0^{(n)})$.
(c)   Rewrite $m=\min(p,q)$ as: $(p\le q\wedge m=p)\vee(q<p\wedge m=q)$.
(e)   Rewrite as $(m=0\wedge n=0)\vee(m\ne0\wedge n=1)$.
4(a)   For $m=0$, $n=0$, show that

$$\vdash_{\mathscr{N}}(0^{(0)}=0\wedge0^{(0)}=0)\vee(0^{(0)}\ne0\wedge0^{(0)}=0^{(1)}).$$

For $m\ne0$, $n=1$, show that

$$\vdash_{\mathscr{N}}(0^{(m)}=0\wedge0^{(1)}=0)\vee(0^{(m)}\ne0\wedge0^{(1)}=0^{(1)}).$$

For $m=0$, $n\ne0$ show that

$$\vdash_{\mathscr{N}}\sim(0^{(0)}=0\wedge0^{(n)}=0)\vee(0^{(0)}\ne0\wedge0^{(n)}=0^{(1)}).$$

For $m\ne0$, $n=0$, show that

$$\vdash_{\mathscr{N}}\sim((0^{(m)}=0\wedge0^{(0)}=0)\vee(0^{(m)}\ne0\wedge0^{(0)}=0^{(1)}).$$

Last, for $m = 0$ and $m \neq 0$ separately

$$\vdash_{\mathscr{N}} (\exists_1 x_1)((0^{(m)} = 0 \wedge x_1 = 0) \vee (0^{(m)} \neq 0 \wedge x_1 = 0^{(1)})).$$

(b) Let $\mathscr{A}(x_1, x_2)$ be the wf. $x_2 = x_1 + 0^{(3)}$.

If $n = m + 3$ then $\vdash_{\mathscr{N}} 0^{(n)} = 0^{(n)} + 0^{(3)}$.

If $n \neq m + 3$ then $\vdash_{\mathscr{N}} \sim (0^{(n)} = 0^{(m)} + 0^{(3)})$.

Also for any $m \in D_N$, $\vdash_{\mathscr{N}} (\exists_1 x_2)(x_2 = 0^{(m)} + 0^{(3)})$.

(Alternatively use Example 6.7).

(c) Let $\mathscr{A}(x_1, x_2)$ be $(\exists x_3)((x_1 = x_2 + x_3 \times 0^{(2)}) \wedge (x_2 = 0 \vee x_2 = 1))$.

5   This is a complicated exercise on the use of the Deduction Theorem. We must show that if $f(m) \neq n$ then $\vdash_{\mathscr{N}} \sim \mathscr{A}(0^{(m)}, 0^{(n)})$. Let $f(m) = p$. We need to prove:

$$\{0^{(n)} \neq 0^{(p)}, \mathscr{A}(0^{(m)}, 0^{(p)}), (\exists_1 x_2)\mathscr{A}(0^{(m)}, x_2)\} \vdash_{\mathscr{N}} \sim \mathscr{A}(0^{(m)}, 0^{(n)}).$$

It is sufficient to show:

$$\{0^{(n)} \neq 0^{(p)}, \mathscr{A}(0^{(m)}, 0^{(p)})\} \vdash_{\mathscr{N}} (\exists_1 x_2)\mathscr{A}(0^{(m)}, x_2) \rightarrow \sim \mathscr{A}(0^{(m)}, 0^{(n)})$$

or

$$\{0^{(n)} \neq 0^{(p)}, \mathscr{A}(0^{(m)}, 0^{(p)})\} \vdash_{\mathscr{N}} \mathscr{A}(0^{(m)}, 0^{(n)}) \rightarrow \sim (\exists_1 x_2)\mathscr{A}(0^{(m)}, x_2)$$

or

$$\{0^{(n)} \neq 0^{(p)}, \mathscr{A}(0^{(m)}, 0^{(p)}), \mathscr{A}(0^{(m)}, 0^{(n)})\} \vdash_{\mathscr{N}} \sim (\exists_1 x_2)\mathscr{A}(0^{(m)}, x_2).$$

Intuitively this is clear, but there are technical difficulties still. We need

$$\{0^{(n)} \neq 0^{(p)}, \mathscr{A}(0^{(m)}, 0^{(p)}), \mathscr{A}(0^{(m)}, 0^{(n)})\} \vdash_{\mathscr{N}}$$

$$(\forall x_2) \sim (\mathscr{A}(0^{(m)}, x_2) \wedge (\forall x_3)(\mathscr{A}(0^{(m)}, x_3) \rightarrow x_3 = x_2)).$$

This last wf. is equivalent to

$$(\forall x_2)(\mathscr{A}(0^{(m)}, x_2) \rightarrow \sim (\forall x_3)(\mathscr{A}(0^{(m)}, x_3) \rightarrow x_3 = x_2)).$$

So we show that

$$\{0^{(n)} \neq 0^{(p)}, \mathscr{A}(0^{(m)}, 0^{(p)}), \mathscr{A}(0^{(m)}, 0^{(n)}), \mathscr{A}(0^{(m)}, x_2)\}$$

$$\vdash_{\mathscr{N}} \sim (\forall x_3)(\mathscr{A}(0^{(m)}, x_3) \rightarrow x_3 = x_2)$$

by showing that

$$\{\mathscr{A}(0^{(m)}, 0^{(p)}), \mathscr{A}(0^{(m)}, 0^{(n)}), \mathscr{A}(0^{(m)}, x_2), (\forall x_3)(\mathscr{A}(0^{(m)}, x_3)$$

$$\to x_3 = x_2)\} \underset{\mathscr{N}}{\vdash} 0^{(n)} = 0^{(p)}.$$

(N.B. We may not use Generalisation on x_2 in this last step.)

6 Let $\mathscr{A}(x_1, \ldots, x_{k+1})$ be $x_{k+1} = x_i$.

Section 6.3 (p. 145)

7(a) $e(m, 0) = 1$

$e(m, n+1) = m \times e(m, n).$

Now $e(m, 0)$ is a constant function of m, and this is primitive recursive. Also $m \times e(m, n)$ is a primitive recursive function of m, n and $e(m, n)$. We may write $m \times e(m, n)$ as $h(m, n, e(m, n))$ where $h(m, n, p) = p_1^3(m, n, p) \times p_3^3(m, n, p)$. The dependence on n is not explicit in $m \times e(m, n)$, and because of the above device, it does not need to be.

(b) $\min(m, n) = m \dot- (m \dot- n).$

(c) First let $rm(m, n) = \begin{cases} \text{remainder on division of } n \text{ by } m \text{ if } m \neq 0 \\ 0 \qquad\qquad\qquad\qquad\qquad\qquad \text{if } m = 0. \end{cases}$

Then $rm(m, 0) = 0$

$rm(m, n+1) = sg(m) \times sg((m \dot- 1) \dot- rm(m, n)) \times (1 + rm(m, n)),$

so rm is primitive recursive, since $+, \times, \dot-,$ and sg are.

Now $q(m, 0) = 0$

$q(m, n+1) = sg(m) \times (q(m, n) + \overline{sg}((m \dot- 1) \dot- rm(m, n))),$

so q is primitive recursive since $+, \times, \dot-, sg, \overline{sg}$ and rm are.

8 R is recursive, so C_R is recursive and $R(n_1, \ldots, n_k, x)$ holds if and only if $C_R(n_1, \ldots, n_k, x) = 0.$

9 $e_2(n) = \mu x[\overline{sg}(rm(2^x, n)) = 0] \dot- 1.$

11 Let $k(n_1, n_2) = n_1^{n_2}$. k is recursive, f and g are recursive so h is recursive, by composition, since

$h(x) = k(f(x), g(x)).$

14 Define one-place relations $R_1, R_2, \ldots$ by:

$R_i(n)$ holds if and only if $R(i, n)$ holds.

Then $S(n)$ holds if and only if $R_1 \vee R_2 \vee \ldots \vee R_{k-1}$ holds, and

$T(n)$ holds if and only if $R_1 \wedge R_2 \wedge \ldots \wedge R_{k-1}$ holds.

With the condition '$< k$' removed, these relations are no longer necessarily recursive. Consider a non-recursive set X, listed as $\{x_1, x_2, \ldots\}$ and let

$X_i = \{x_1, \ldots, x_i\}$, for $i \geq 1$. Then let $R(m, n)$ hold if and only if $n \in X_m$. The R_i defined as above are recursive, since the X_i are finite sets, but $S^\infty(n)$ is not recursive, where $S^\infty(n)$ holds if and only if there is m such that $R(m, n)$, since then $S^\infty(n)$ holds if and only if $n \in X$, and X is not recursive.

Section 6.4 (p. 150)

1 5(a) $65 = 9 + 8 \times 7 = g(a_7)$.
 (b) $299 = 11 + 8 \times 36 = 11 + 8 \times (2^2 \times 3^2) = g(f_2^2)$.
 (c) $109 = 13 + 8 \times 12 = 13 + 8 \times (2^2 \times 3) = g(A_1^2)$.
 (d) $421 = 13 + 8 \times 51 \neq g(t)$ for any symbol t.

16(a) $A_1^1(x_1)$.
 (b) $\sim A_1^1(x_1)$.
 (c) $(\forall x_1)A_1^1(x_1)$.

17   If $s = a_1, a_2, \ldots, a_u$ and $t = b_1, \ldots, b_v$, then $m = p_1^{a_1} \times \ldots \times p_u^{a_u}$ and $n = p_1^{b_1} \times \ldots \times p_v^{b_v}$. Now

$$s * t = a_1, a_2, \ldots, a_u, b_1, \ldots, b_v,$$

so

$$f(m, n) = p_1^{a_1} \times \ldots \times p_u^{a_u} \times p_{u+1}^{b_1} \times \ldots \times p_{u+v}^{b_v}.$$

But $b_i = e_i(n)$, $u = \mu x [\overline{sg}(rm(p_x, m)) = 0] \dotminus 1$ and since p_x is a recursive function of x (see Example 6.25(a)) u is a recursive function of m. Similarly v is a recursive function of n, so

$$f(m, n) = m \times p_{u+1}^{e_1(n)} \times \ldots \times p_{u+v}^{e_v(n)} \text{ is recursive.}$$

Section 6.5 (p. 155)

18   Denote the extension by $\mathcal{N}^+$. Then $\vdash_{\mathcal{N}^+} \sim \mathcal{U}$,

i.e. $\vdash_{\mathcal{N}^+} \sim (\forall x_2) \sim \mathcal{W}(0^{(p)}, x_2)$.

But in the proof of Proposition 6.32 it is shown that $\vdash_{\mathcal{N}} \sim \mathcal{W}(0^{(p)}, 0^{(q)})$ for every q. Hence $\vdash_{\mathcal{N}^+} \sim \mathcal{W}(0^{(p)}, 0^{(q)})$ for every q, and $\mathcal{N}^+$ is not ω-consistent.

Chapter 7

Section 7.1 (pp. 163-4)

1(a) In Chapter 6 we describe methods for finding what symbols or strings of symbols corresponded with given Gödel numbers. Given n, find what symbol (if any) it corresponds to. If it is a variable or a constant then give answer Yes. If it is another symbol then answer No. If there is no symbol, then find what string of symbols it corresponds to. If it is a term then answer Yes. If it is another string or there is no string, answer No.

(c) Calculate the squares of all positive integers until we find $x^2 \geq n$. If n has not appeared in the list by this stage then it is not a perfect square.

(d) Construct truth tables for all of the wfs. in Γ and for $\mathcal{A}$. $\vdash \mathcal{A}$ if and only if whenever all wfs. of Γ are T, $\mathcal{A}$ is also T.

(f) Make a list of all primes by checking each positive integer in turn for primeness and putting only the primes in the list.

2   Let A and $\bar{A}$ be recursively enumerable. Then let $f(0)$, $f(1)$, $f(2)$, ... and $g(0)$, $g(1)$, $g(2)$, ... be effective enumerations of A and $\bar{A}$. Let $n \in D_N$. Write out (simultaneously) the lists above for A and $\bar{A}$. n must occur in one of the lists. When it does occur, the list it is in tells us whether or not $n \in A$ or $n \in \bar{A}$. There is a set X which is not recursive. Then $\bar{X}$ is not recursive (Corollary 6.23). One at least of X and $\bar{X}$ is not recurisvely enumerable, since if both were then both would be recursive.

3   Given x, enumerate A until the first element $\geq x$ is found. $x \in A$ if and only if x is in this list.

4(a) Use Church's Thesis. Given n, for each $p < n$ find whether p and n have any common factor. Count those p which have. This is an algorithm for computing ϕ.

(c) Either there are sequences of 7s of arbitrary length, or there is an upper bound, say k, to the lengths of such sequences. In the first case $g(n) = 0$ for all $n \in D_N$ (so g is recursive). In the second case

$$g(n) = \begin{cases} 0 & \text{for } 0 \leq n \leq k \\ 1 & \text{for } n > k \end{cases}$$

and this function is recursive also.

5(a) D_N.

(b) The set of Gödel numbers of theorems of $\mathcal{N}$. The set of Gödel numbers of wfs. of $\mathcal{N}$ which are tautologies is a recursive subset.

6   See Exercise 3. Given an effective enumeration of a set A, pick out an *increasing* enumeration of a subset B. B will be recursive.

Section 7.2 (pp. 182–3)

7   The initial tape will need a marker at the right hand end of the non-blank part, say M. Add the quadruples $(q_0 \ B \ R \ q_0)$ and $(q_0 \ M \ B \ q_2)$.

9   $\{(q_0 \ B \ R \ q_0), (q_0 \ 1 \ R \ q_0)\}$

or

$\{(q_0 \ B \ R \ q_1), (q_0 \ 1 \ R \ q_1), (q_1 \ B \ L \ q_0), (q_1 \ 1 \ L \ q_0)\}$.

10 $\{(q_0 \ 1 \ A \ q_1), (q_1 \ A \ R \ q_2), (q_2 \ 1 \ R \ q_2), (q_2 \ B \ X \ q_2),$
     $(q_2 \ X \ R \ q_3), (q_3 \ B \ 1 \ q_4), (q_3 \ 1 \ R \ q_3), (q_4 \ 1 \ L \ q_4),$
     $(q_4 \ X \ L \ q_4), (q_4 \ A \ R \ q_0), (q_0 \ X \ L \ q_5), (q_5 \ A \ 1 \ q_5), (q_5 \ 1 \ L \ q_5)\}$
     Machine starts in state q_0 reading leftmost 1.

11 See Example 7.13. $\{(q_0\ 1\ R\ q_1),\ (q_1\ 1\ R\ q_0),\ (q_0\ B\ R\ q_0)\}$ (if the input number is even, the quadruple $(q_0\ B\ R\ q_0)$ ensures that the machine never stops). To obtain T' omit $(q_0\ B\ R\ q_0)$ and add the quadruples $(q_0\ B\ L\ q_2)$, $(q_0\ 1\ B\ q_3)$, $(q_3\ B\ L\ q_2)$. Consider the list $\phi_0, \phi_1, \ldots$ of all recursive partial functions. Then $f(n) = \phi_n(n) + 1$ is a recursive *partial* function which cannot be extended to a recursive *total* function. For suppose ϕ_k is total and $\phi_k(n) = \phi_n(n) + 1$ wherever $\phi_n(n)$ exists. Now ϕ_k is total, so $\phi_k(k)$ exists, and $\phi_k(k) = \phi_k(k) + 1$. Contradiction. (See the example before Proposition 7.28.)

12 Suppose the alphabet of tape symbols contains n symbols, and there are k internal states. If the machine does not move from a given square in $nk + 1$ steps then it will never halt, for in that time it must repeat a (state, symbol being read) pair and from then it repeats a periodic action for ever. (Of course it may halt before $nk + 1$ steps.) If the non-blank part of the tape originally consists of p squares, then after $p(nk + 1)$ steps we can be sure that the machine will have halted, or moved off the right hand end of the non-blank tape, or have entered a 'stationary' loop as above. If it moves off the right hand end of the non-blank tape then it can move at most k further squares right without repeating a (state, blank) combination. Thus if it moves $k + 1$ further squares then it must enter a repetitive pattern. Now after $(k + 1)(nk + 1)$ further steps it will have halted or entered a stationary loop or have moved $k + 1$ further squares right. We can therefore say in advance that if the machine is going to halt at all it will do so in $(p + k + 1)(nk + 1)$ steps or fewer, so the algorithm consists of running the machine until it stops or for $(p + k + 1)(nk + 1)$ steps.

15 Suppose that $n \in K$. Then T_n halts with input n, so $n \in A$. But $A \subseteq \bar{K}$, so $n \in \bar{K}$. Contradiction, so $n \in \bar{K}$. Now suppose that $n \in A$. Then T_n halts with input n, so $n \in K$. Contradiction, so $n \notin A$. Hence $n \in \bar{K} \setminus A$.

16 Let X be recursively enumerable. Suppose that X is enumerated by the recursive function f. Define g by

$$g(y) = \mu x[f(x) = y].$$

Because f is recursive, g is a recursive partial function, and its domain is X. By Proposition 7.25, g is Turing computable, and so there is a Turing machine with domain X.

17(e) $\{$does T_n halt for every input number? $\mid n \in D_N\}$. Consider the algorithm: Let m be fixed. Given n, follow the computation of T_m, with input m; and if it halts, give output n. This algorithm can be transformed first into instructions for a Turing machine and second into a code number $k(m)$ for a Turing machine $T_{k(m)}$. $T_{k(m)}$ has the property that if T_m halts with input m then $T_{k(m)}$ halts for every input, and if T_m does not halt with input m then $T_{k(m)}$ halts for no input. Now to decide whether T_m halts with input m we need only calculate $k(m)$ and answer the question: does $T_{k(m)}$ halt for every input? Thus solvability of the class of problems posed would imply tht K is recursive.

(f) $\{$does T_n halt for some input number? $\mid n \in D_N\}$.

$T_{k(m)}$ (in (e) above) either halts for no inputs or halts for every input, depending on m. Thus if we can decide whether $T_{k(m)}$ halts for some input number, then we can decide membership of K, as in (e).

18   An algorithm for deciding membership of K_0 would clearly yield an algorithm for deciding membership of K, so K is reducible to K_0. Now suppose an algorithm exists for K. Consider the algorithm: Let m, n be fixed. Given p, follow the computation of T_m with input n; and if it halts, give output p. As in Exercise 17(e) this algorithm yields instructions for a Turing machine and hence a code number $k(m, n)$ for this machine, and $T_{k(m,n)}$ halts for every input p or for no input, depending on whether T_m halts with input n. Thus to decide whether $(m, n) \in K_0$ we need only calculate $k(m, n)$ and answer the question: does $T_{k(m,n)}$ halt with input $k(m, n)$? i.e. is $k(m, n) \in K$?

Section 7.3 (p. 189)

19(a)   Any given word may be easily shown to be equivalent to a word in one of the following standard forms:

$$a_1 a_2^{k_1} a_1 a_2^{k_2} a_1 \dots a_1 a_2^{k_n} a_1 \qquad (n \ge 1),$$

$$a_1^{k_1} a_1 a_2^{k_2} a_1 \dots a_1 a_2^{k_n} a_1 \qquad (n \ge 0),$$

$$a_2^{k_1} a_1 a_2^{k_2} a_1 \dots a_1 a_2^{k_n} \qquad (n \ge 1).$$

Two words in standard form are equivalent if and only if they are identical, so two given words are equivalent if and only if they reduce as above to the *same* standard form.

(b)   Note that $a_3 a_2 \sim a_1 a_2 a_2 \sim a_1 a_2 \sim a_3$.

Any given word may be shown to be equivalent to one in the form

$$a_1^{k_1} a_3^{k_2} a_1^{k_3} \dots a_1^{k_{n-1}} a_3^{k_n},$$

where all of the occurrences of a_2 have been 'absorbed'.

20   G is an abelian group. That the symbols $a_1^{-1}, a_2^{-1}, a_3^{-1}$ commute with the other symbols and with each other is an exercise in manipulation. For example:

$$e \sim a_2^{-1} a_1^{-1} a_1 a_2 \Rightarrow e \sim a_2^{-1} a_1^{-1} a_2 a_1$$

$$\Rightarrow e a_1^{-1} \sim a_2^{-1} a_1^{-1} a_2 a_1 a_1^{-1} \sim a_2^{-1} a_1^{-1} a_2 \Rightarrow a_1^{-1} a_2^{-1} \sim a_2^{-1} a_1^{-1},$$

and

$$a_1 a_2^{-1} a_1^{-1} a_2 \sim a_1 a_1^{-1} a_2^{-1} a_2 \sim e,$$

and so

$$a_1 a_2^{-1} a_1^{-1} a_2 a_2^{-1} \sim a_2^{-1}$$

and

$$a_1 a_2^{-1} a_1^{-1} a_1 \sim a_2^{-1} a_1,$$

i.e.

$$a_1 a_2^{-1} \sim a_2^{-1} a_1.$$

Any given word may be reduced to a standard form $a_1^{r_1} a_2^{r_2} a_3^{r_3}$, where r_1, r_2, $r_3 \in \mathbb{Z}$. Two given words are equivalent if and only if their standard forms are identical.

22 Just as any Turing machine can be replaced by an equivalent one which uses just two tape symbols (Remark 7.22), so we can construct essentially the same semigroup $\mathscr{S}$ in Theorem 7.37 using only two symbols. The symbols of $\mathscr{S}$ can be coded in unary form, with a 0 at the beginning to act as a marker. E.g. q_0 would be 01, q_4 would be 011111 and the word $q_0 q_4 q_1$ would be 01011111011.

Section 7.4 (pp. 197–8)

24(a) $\{n \in D_N \colon n$ is the Gödel number of $\sim \mathscr{A}$, where $\mathscr{A}$ is a theorem of $\mathcal{N}\}$ is not recursive. Given $n \in D_N$ find whether it is the Gödel number of a *wf*. $\mathscr{A}$ of $\mathcal{N}$. If it is, calculate the Gödel number of $(\sim \mathscr{A})$ (this can be done effectively – see Section 6.4). If the given set were recursive then the set of Gödel numbers of theorems of $\mathcal{N}$ would be recursive.

(b) $\{n \in D_N \colon n$ is the Gödel number of $\mathscr{A}$, where $\mathscr{A}$ is true in $N\}$ is not recursive (Ch. 6). $\mathscr{A}$ is true in N if and only if $(\sim \mathscr{A})$ is false. Thus recursiveness of the given set would yield a contradiction.

26 If $\mathscr{V}$ is true in N then $F(p, s)$ holds for some $s \in D_N$, and so $f(s)$ is the Gödel number of $(\exists x_2)\mathscr{F}(0^{(p)}, x_2)$ for some s, so $\mathscr{V}$ is false in N. $\mathscr{V}$ is closed, and so must be false in N. If $q = f(k)$ (q is the Gödel number of $\mathscr{V}$) then $f(k)$ is the Gödel number of $(\exists x_2)\mathscr{F}(0^{(p)}, x_2)$ so $F(p, k)$ holds. Hence $\vdash_{\mathcal{N}} \mathscr{F}(0^{(p)}, 0^{(k)})$, and so $\vdash_{\mathcal{N}} (\exists x_2)F(0^{(p)}, x_2)$, i.e. $\vdash_{\mathcal{N}} \mathscr{V}$, and so $\mathscr{V}$ is true in N. Contradiction, so q is not in the range of f.

27 If the set of (Gödel numbers of) axioms of T is recursive, then the set of (Gödel numbers of) theorems of T is recursively enumerable (Remark 7.6). The systems S and T have the same set of theorems. If S is complete and its set of (Gödel numbers of) theorems is recursively enumerable, then given any $n \in D_N$, if it is the Gödel number of a *wf*. $\mathscr{A}$ of S, then enumerate all the (Gödel numbers of) theorems of S. Either $\mathscr{A}$ or $\sim \mathscr{A}$ is a theorem of S, and we will eventually find which. Hence S is recursively decidable.

References and further reading

COHEN, P. J, *Set Theory and the Continuum Hypothesis*, Addison-Wesley, 1966.

COPI, I. M, *Introduction to Logic*, Macmillan, 1961.

DAVIS, M (1), Hilbert's tenth problem is unsolvable, *American Mathematical Monthly*, Vol. 80 (1973) p. 233.

DAVIS, M (2), *Computability and Unsolvability*, McGraw-Hill, 1958.

HALMOS, P. R, *Naive Set Theory*, Van Nostrand, 1960.

HILBERT, D, Mathematical problems, *Bulletin of the American Mathematical Society*, Vol 8 (1901–2) p. 437.

KLEENE, S. C, *Introduction to Metamathematics*, Van Nostrand, 1952.

MENDELSON, E, *Introduction to Mathematical Logic*, Van Nostrand, 1964.

MINSKY, M, *Computation: Finite and Infinite Machines*, Prentice-Hall, 1967.

ROBINSON, A, *Introduction to Model Theory and to the Metamathematics of Algebra*, North-Holland, 1965.

ROGERS, H, *Introduction to the Theory of Recursive Functions and Effective Computability*, McGraw-Hill, 1967.

SHOENFIELD, J. R, *Mathematical Logic*, Addison-Wesley, 1967.

TARSKI, A, MOSTOWSKI, A, AND ROBINSON, R. M, *Undecidable Theories*, North-Holland, 1953.

VAN HEIJENOORT, J, *From Frege to Gödel: A Source Book in Mathematical Logic 1879–1931*, Harvard University Press, 1967.

Glossary of symbols

The page number given is that where the symbol is either defined or first used. Page numbers are not given for standard mathematical symbols which are used frequently. The symbols are grouped as follows: English letters, Greek letters, mathematical symbols, logical symbols.

Σ_n-form	prenex form	90
$\varphi_0, \varphi_1, \varphi_2, \ldots$	enumeration of recursive partial functions	178
$\triangleright$	resumption of the main exposition after it has been broken by a proposition, example, remark, corollary or definition	
$\varnothing$	the empty set	121
$\circ$	composition of functions	199
$\dot{-}$	modified subtraction	141
$'$	successor	117
$\wedge$	conjunction	4
$\wedge$	conjunction in $\mathscr{L}$	53
$\vee$	disjunction	5
$\vee$	disjunction in $\mathscr{L}$	53
$\sim$	negation	4
$\rightarrow$	conditional	5
$\leftrightarrow$	biconditional	6
$\leftrightarrow$	biconditional in $\mathscr{L}$	80
$\downarrow$	nor	20
$\mid$	nand	20
$\bigwedge_{i=1}^{n}$	conjunction	14
$\bigvee_{i=1}^{n}$	disjunction	14
$\vdash_{L}$	yields in L	30
$\vdash_{K}$	yields in K	74
$\models$	truth in an interpretation	63
$\forall$	universal quantifier	47
$\exists$	existential quantifier	47
$\exists$	existential quantifier in $\mathscr{L}$	53
$\exists_1$	modified existential quantifier	111
$0^{(n)}$	term in $\mathscr{N}$	130

Index

Printed in the United States
by Baker & Taylor Publisher Services